WE'LL CALL YOU
IF WE NEED YOU

Also by Susan Eisenberg:

It's a Good Thing I'm Not Macho

Pioneering: Poems from the Construction Site

WE'LL CALL YOU
IF WE NEED YOU

Experiences of Women
Working Construction

Susan Eisenberg

ILR Press • an imprint of Cornell University Press
Ithaca and London

First published 1998 by Cornell University Press.

Printed in the United States of America.

Library of Congress Cataloging-in-Publication Data
Eisenberg, Susan.
    We'll call you if we need you : experiences of women working
construction / Susan Eisenberg.
        p.   cm.
    ISBN 0-8014-3360-6 (cloth : alk. paper)
    1. Women construction workers—United States—Interviews.
I. Title.
HD6073.B92U63   1998
331.4′824′0973—dc21   97-45749

Cornell University Press strives to utilize environmentally responsible suppliers and materials to the fullest extent possible in the publishing of its books. Such materials include vegetable-based, low-VOC inks and acid-free papers that are also either recycled, totally chlorine-free, or partly composed of nonwood fibers.

Cloth printing   10 9 8 7 6 5 4 3 2 1

for tradeswomen
past, present, future

and for Zoe and Simon

Contents

The Tradeswomen

Acknowledgments

For the flesh and blood of this book (as well as many nuts and bolts—taking time from busy schedules and life crises to return phone calls and mailings or hunt for photographs), I deeply thank the women who entrusted their stories and perspectives to me. I have tried to be carefully respectful of the gifts given, and apologize for wherever I may have been mistaken.

A combination of small grants and generous hospitality made it financially possible to gather these stories in person. For that I thank the Women's Studies Department of Northeastern University for my appointment as International Research Associate in Women's Studies; the Money for Women/Barbara Deming Memorial Fund; the Samuel Gompers Union Leadership Award of City University of New York; the Ms. Foundation for Women; the Massachusetts Foundation for the Humanities; the American Association of University Women; Tricia Coley and Jim Coley (Seattle); Molly Martin (Bay Area); the George Meany Center for Labor Studies, the Labor Heritage Foundation, and Virginia DuRivage (Washington, D.C.); Judy Ancel and Kathy Walsh (Kansas City); Anne Brophy (Los Angeles); Brenda Bishop and Cynthia Long (New York City); the AFL–CIO Western Regional Summer Institute for Union Women (Olympia and Vancouver); and Bee and Al Eisenberg (Cleveland). For other interview assistance I thank Hard Hatted Women of Cleveland, Tradeswomen, Inc., of California, the Association for Union Democracy, Wider Opportunities for Women, the Boston Tradeswomen's Network, Verlet Allen, Naomi Friedman, and Mary Baird.

For research assistance and advice I thank Arleen D. Winfield of the Women's Bureau/U.S. Department of Labor, Peggy Crull of the New York City Commission on Human Rights, Eva Moseley of the Schlesinger Library, Elaine Bernard, Roslyn Feldberg, Chris Tilly, Brigid O'Farrell, Victoria Fortino, and Cindy Morano. For technical assistance and advice, I thank Eileen Bolinsky, Paul McGowan, Jeff Perkell, Paul Wesel and my endlessly resourceful friend Jack Fahey for help viewing photographs. For contact advice I thank Archie Brodsky and the National Writers Union. Because of a serious hand injury, I am grateful that most interviews were transcribed by Mercury Manuscripts and Word Pros.

Although I take responsibility for whatever errors remain, I thank for their careful reading of the entire manuscript Karen Pollak, Maura Russell, Gay Wilkinson, Mary Baird, and the meticulous Virginia DuRivage. Larry Levis, Alan Williamson, Vivian Troen, Barbara Gottlieb, and Charles Coe gave feedback on individual chapters. Thanks to the Boston Tradeswomen's Network for the chance to hear the "Passions" chapter aloud. The structural concept took first form in a workshop with Lady Borton at the William Joiner Center and evolved in the greenhouse of the Master of Fine Arts Program for Writers at Warren Wilson College; special thanks to Eleanor Wilner for her humor, support, and standards taught by example.

I doubt that this book could have come to publication without the path cleared for it by Fran Benson, editor-in-chief at Cornell University Press, who believed in it from the start and encouraged me to tell it my own way. I thank her and her assistant Laura Healy for their enthusiasm, trust, and availability. I am grateful to the entire team at the Press, in their various departments, for the responsive and collaborative effort given this book.

For my own story, I thank the International Brotherhood of Electrical Workers and the many exceptional men and women in that union, and the grassroots community of tradeswomen across the country and in Canada who have helped me laugh and think. For telling me a story that inspired this book, I thank Frances Knowles.

S. E.

WE'LL CALL YOU
IF WE NEED YOU

Thumbnail Sketch

Working the steel and seeing a building
rise from the ground
is right up there with giving birth.
To know that your quality of work
ensures the safety of so many people in a high-rise
gives a feeling of satisfaction unlike any other.
—*Gay Wilkinson*

I had a good motivating force to try it,
because there was a lady in the office,
nice little short lady. She was stooped over,
had one of those little curved hunches in her back.
I took a good look at her one day and I said to myself,
I don't want to be like that twenty or thirty years from now.
—*Gloria Flowers*

They see that women can do it, they see that women
are good at it—I had this expectation that barriers
would just fall away. I mean in the first few years
it was increasing. At least five, six women per class.
One year there were ten or something. It was like–
whoa! Daddy! This is very exciting!
—*Sara Driscoll*

Is my desire to change this industry
so that it's not such a hostile place for women
worth it? Is it worth my life? Is it worth being crippled?
Is it worth all the psychological harm? Sometimes you say,
yes, it is. If I don't do it, who *is* going to do it?

—*Cynthia Long*

Maybe the job that they got coming up, I can do
a little better than you. Do you ever think of that?
That we may be equal to doing work
but there might be one more thing that I do
better than you and I got kept because I can do that.
Maybe I can tape and paint and you can only paint.
So they're keeping me because they can use me more,
not the fact because I'm a girl. If you sit down
and sift through it you might find the real reason
why I got kept and not you.

—*Deb Williams*

Introduction

When I began my electrical apprenticeship in 1978 with Local 103 of the International Brotherhood of Electrical Workers, I felt very lucky. I was going to learn a trade that would challenge me both physically and intellectually and would earn me a solid income. I would have the protection and camaraderie of belonging to a labor union. I would be part of a historic social and economic effort to open higher-paying skilled trades jobs to women—initiated and supported by the federal government, in recognition both of women's economic responsibilities and of the full range of our capabilities.

In April 1978, the Department of Labor under President Jimmy Carter set hiring goals and timetables for hiring women on federally funded construction projects on a trade-by-trade basis. These goals were to increase over a three-year period to 6.9 percent of the workforce. Additional regulations established that women should make up between a fifth and a fourth of each apprenticeship class. I just assumed that these guidelines which had opened doors for me would continue to be enforced and to increase the number of women in the industry.

The regulations established in 1978 blueprinted a course for the construction workforce just past the start of the millennium to become about one quarter female, making the occupation no longer a nontraditional one for women. This is not what happened. Women's overall percentage of the construction workforce grew to roughly two percent in the early 1980s and remained there.

As a new apprentice I also believed that the many capable tradeswomen who came into the industry would, by force of our positive example, disprove the notion that construction was by nature "men's work." I believed that our accomplishments would inspire the gatekeepers to welcome other women, to see the *absence* of women in the skilled trades as illogical, archaic, silly—an inequity to correct.

But they didn't. Despite many examples of tradeswomen who were skilled, conscientious, and productive workers, despite many efforts by tradeswomen as individuals and through grassroots organizations to expand the participation and retention of women, the culture and the workforce composition of the traditionally male industry have remained fairly intact. It seemed that a tradeswoman's performance could become cause for other women's exclusion, but not for their inclusion. The myth that *no* women were capable of skilled trades work was replaced by the Myth of the Exceptional Woman, the disconnection of any successful woman from the rest of her gender.

The industry has made adjustments in response to external factors such as new laws governing sexual harassment and pregnancy disability in the workplace. And a token percentage of women have been able to build financially successful and satisfying careers in the skilled trades. But the promise created by Executive Order 11246, that access to these higher paid blue-collar jobs would create a viable career option for a significant number of women, has not been realized. For women, the pioneering phase of breaking into union construction was not followed by a critical mass of settlers. Instead, pioneering, contrary to its meaning, became a seemingly permanent condition.

Such an enormous disparity between goal and result requires explanation. I found the most common of these very troubling: that once women realized the work was difficult or dirty, they decided it wasn't for them and left. That women are not capable of the work. That although an opportunity was offered to women on a silver platter, there was no interest. These answers matched neither my own experience nor what I had heard from other tradeswomen through countless conversations on jobsites, at conferences, or when I'd traveled with my poetry.

I began in 1991 to gather oral histories of thirty women from ten states and the District of Columbia who were among the first in their union locals in five trades— carpenters, electricians, ironworkers, painters, and plumbers. A third of the group are women of color. One left the trades after two months, while another became senior business representative of her local. By documenting and considering the experiences of other tradeswomen, I hoped to better understand what happened to me and the moment of history I have been part of—especially why there weren't more of us.

The first interview was at Thanksgiving, 1991. It's been a long process that has depended on the help of a great many people: grassroots tradeswomen's organizations who helped me locate tradeswomen in their area; individuals who picked me up at an airport, put me on a bus, or welcomed me into their homes; grantors who sometimes widened their guidelines to give this project support.

But most of all I am deeply grateful to the women I interviewed, who generously undertook the sometimes difficult emotional journey of bringing me back with them to the beginnings of their careers in construction. I am indebted to those women and to many others in the tradeswomen community for the countless conversations, the shared ideas, and the encouragement that were crucial to bringing this book to completion.

I knew from the outset that I wanted this book to convey the perceptions that tradeswomen share with each other. What has always struck me as problematic is the lack of working tradeswomen at the policy table. I asked every woman I interviewed, given that the entrance of women was a brand-new experience for the industry, was your advice on how to incorporate women requested? Not one said yes. So the book ends with suggestions, less definitive than the punch list of uncompleted tasks handed to a contractor at the end of a job. But I hope that those thoughts, and the book itself, will contribute to a conversation about what would be required to make a historically male industry truly as hospitable to woman as it is to men. Since I believe that conversation belongs in the public arena, I have tried to write the book to be understandable even by readers with no experience at all in the trades.

Rather than organizing the book so that each chapter is about a different woman, I arranged it thematically, beginning with women's decisions to enter the industry. The main character essentially becomes the movement of tradeswomen as a whole through the industry—not always of one mind, as one sometimes even argues with oneself. I hope this arrangement helps the reader to resist the temptation simply to glorify and romanticize these women, and to focus instead on identifying the many points at which the progressive agenda became derailed, and on imagining solutions. Rereading the 1978 federal guidelines in detail, I was surprised at their forethought and how well they did address—on paper—many of the problems to come.

It is particularly important to hear the perspectives of these women now, when "welfare reform" again places high on the public agenda the issue of how women will earn a living wage for themselves and their families. It is also important, at this moment in the global economy when the workforce is becoming increasingly temporary, to examine this industry where work has always been temporary, to see how cycles of layoffs and hirings can serve to mask discrimination by gender and race.

In the interviews and in editing and arranging them for this book, I tried to draw on my training in the trades, in women's studies, and, most of all, in poetry. I tried to hear the significance of what each woman was saying within the context of the industry, the social and economic historical moment, and her own personal life, which carries imprints from all of that. When our differences made it hard for me to grasp what they were trying to tell, women were extraordinarily patient.

In translating from a transcription of oral language to a more readable written format, I have tried to keep in mind what I learned from Denise Levertov about listening to the human thought process. I noticed how often in telling a story a woman would weave back and add in another thread, or rework her language. I have tried to keep in some of those digressions, and when editing down, to select the last choice a person made in phrasing. I have occasionally rearranged the order of sentences for clarity. In paring down, I have been aware that the eye is less tolerant to words on a page than the ear is to a live human voice. I have tried to maintain in written form the energy and integrity of spoken language.

Of the thirty women interviewed, twenty-eight appear in the book. One woman chose not to appear; another, painter Rae Sovereign, I chose not to include because she always worked maintenance, not construction. I thank them both for widening my understanding. All but one woman has chosen to be identified by her own name.

Although every woman I spoke to humbled me by her courage and intelligence, I believe any thirty women from those trades and those times would also have amazing stories to tell and thoughtful perceptions to share. Other than fitting specific criteria, the process for selecting the women I interviewed was essentially a random one. Likewise, in the chapter on Exceptional Men, each one stands for many. Any individuals, union locals, or contractors mentioned are also emblematic of situations occurring at the time. Some names are indicated with only a letter, to emphasize that this book's intention is not to paint villains or heroes (or heroines), but to identify the framework in which this on-going story of women's journey through the construction site gate takes place. I hope the reader will understand as well why some women have chosen to stay in the industry as why some have chosen to leave, and how very complex this issue is.

As I write this introduction, the IBEW is planning its first-ever national women's conference, in September 1997. I hope that this book adds to the discussion beginning to percolate in so many arenas about what it would take from all of us to remove gender barriers from the workplace. Construction is a publicly visible and inherently dramatic industry, but the problems women have faced there will, I suspect, find a resonance of familiarity even with women who work in fields and under conditions that are starkly different.

CHAPTER ONE

Footings

> This friend of mine worked for the Missouri Unemployment Office
> and he said—this would have been the fall of '77—"They've got
> apprenticeship programs. Go down and check it out. They're start-
> ing affirmative action and this thing is going to open wide up."
>
> —*Diana Suckiel, Kansas City, Missouri*

— *Paulette Jourdan*   (*born 1943*) —

The first time I thought I could make something, I was a kid. All I did was have a dream about building a cage, because we lived in a place where we had lots of squirrels and I wanted to catch a squirrel. I just was going to make it with sticks and string, but I think I was a little afraid of the squirrels. But that's the dream I always remember. I was maybe nine. Then I got caught up in life.

My father was a printer, a lithographer, and my mother never worked. They were very religious, fundamentalist religious, and he didn't ever want her to work. There were six kids and they raised us sort of like, you're *in* this world, but you're not *of* this world. We were very sheltered. I didn't know anything. We were never allowed to go anywhere. We had very few friends.

I graduated high school '61 and went straight to college. I didn't know what I wanted to do, so I dropped out and got married and started having kids in '65. I was married eight years, got divorced, and it was very hard to survive. That's when I started wondering about the trades. I was just very tired of falling into jobs. They were mostly bookkeeping jobs with large companies, banks, small companies. I just thought, I'd like to work with my hands. It was 1975. I knew what carpentry was, but at that time the starting pay for apprentices was like three and a half bucks an hour. I'm thinking, I'm making at least $5 or $6. I can't take a cut in pay like that, so I let it go.

The last job I had was with a very small company, it was partial bookkeeping, credit, receivables, payables—nothing that required a lot of heavy skills. But my boss left, and the woman who replaced her started getting rid of people. Pretty soon, I'm out of a job, and I had no notion of how I was going to find another one. I have a very dear friend, and she remembered that I thought I might want to get into the trades. She heard on the radio the Plumbers' exam was open. She called up and said, "You better get over there and find out." It was 1980. I thought, Plumber. What's a plumber? I had no notion, none.

I had heard the term "construction worker," and I had heard awful things said about them, but I'm not really sure if I even knew what a construction worker was. I was just blotto, I was just finding out what the world is about. I think the only tool I ever touched was maybe a hammer, a screwdriver, and a pair of pliers in my entire life. So I didn't even have a clear notion of what plumbers did. I think I knew they fixed faucets, but I didn't know anything else, nothing. I didn't know pipes were in the walls, I didn't know they were under the house. Go to the bathroom, flush the toilet, I never thought about it. Luckily, the exam had nothing to do with the trades.

— *Randy Loomans* (born 1951) —

I graduated from high school in 1969, worked at a Bausch & Lomb making eyeglasses for doctors. I became pregnant in 1972 and had a daughter, Danette, and spent most of those years not working. Because my husband didn't want me to—he made a good enough income that I didn't have to work. He was a salesman of interiors—carpets, formica, vinyl—and a store manager.

But I always wanted to work, and it was kind of a struggle between me and him. After Danette was pre-school age, I decided I wanted to go to work even part-time, so I worked at Sears while she went to pre-school. Usually three days a week and some weekends. In 1981 we divorced. A lot of it had to do with the issue of working, because all my young life I had worked. I was raised on a 600-acre farm my dad ran. We dug potatoes. We did sugar beets. We raised livestock. We moved pipeline every day. Work was a part of our life. We lived so far out in the country that when we were in fifth and sixth grade we'd have to drive a car to the bus stop to get us to school. I would drive usually because I was the oldest girl. I learned to drive every kind of equipment. Tractors, everything you can imagine.

The actual traditional housewife role didn't fit my needs. My daughter was eight years old when we separated. At that time, the job market in Seattle was real bad. I looked for a job forever, and I finally landed one at a metal-plating company for $4.83 an hour. And it was really a laborious hard job. It was a union shop, but you couldn't join the union until after you'd been there thirty days. What they did was— when they got behind they'd bring in twelve new employees, work you for twenty-nine days, lay you off. That was my first interaction with unions and how they worked. Even at that, the top women in there were only making $8 an hour and had been there for twenty years.

I just thought, Uh. I was used to living a real decent life. I lived it with my husband and I thought, If he can do this for us, I can do it for me. The operators that did the plating were making $12 an hour and I thought, If I stay here, I'm going to do that. I'll go that far. I'll never just make $8 an hour. Then they laid us off, then they'd hire us back—so we could never join the union.

In the meantime, one of my friends was in the Ironworkers and had said they needed women to join the apprenticeship. I went down.

— Cynthia Long　　　　　　(born 1955) —

I was raised in an area called La Tuque, Quebec, in Canada. We moved down here in, I would say, '64, '65, 'cause that's when the Civil Rights Act came into being, and it also changed some immigration restrictions established in the 1800s. Up until 1965, Asians, specifically Chinese women, were excluded from immigration. Basically, they were willing to take Asian or Chinese men to build the railroads, but they didn't want them to establish families, so they kept the women from coming in.

For many years my parents were applying to immigrate into the United States and it was not possible, because they had quotas in terms of how many. Finally, because we were able to meet the criteria, that is, we had family here in the United States and my father is a professional, a mechanical engineer, we were able to come down to New York. 'Cause my grandmother was here. She had the fortune to be born in the United States, which made her an automatic citizen, even though she in fact spent most of her years in China, in Shanghai. So I would have to say I'm first-generation, technically.

I went to college at State University of New York at Buffalo between 1973 and 1975. At the time I couldn't have told you why I left. Now, I am able to say that the reason that I dropped out of college was because I experienced date rape and I psychologically was unable to cope with that and ended up having to leave. It was quite an extreme disappointment to me personally, as well as feeling embarrassed and ashamed that I flunked out of college.

So around 1975 I started to try to look at, What jobs *could* I get, what jobs did I *want* to get? And the kinds of jobs I was able to get was like, file clerk. From that work experience I realized, Yes, I could file and do a good job at it, but it was very emotionally unsatisfying, there was no career ladder, there was no future. That's when I started to, in a more intellectual and analytical way, look at, Okay, what kinds of careers are there? What can I do without a college degree? And that's also when I started to recognize that people can be intimidated by their perception of you and decide not to hire you. I've had many people who look at me, they see an Asian woman and their automatic assumption is I am a foreigner. Their expectation is someone who is not educated, someone who has broken English. So they talk to you really loud, like you're deaf. And slow, like you're retarded. Then when you speak they're like, Oh my God, how do you know English so well? It's like, Why shouldn't I know English? English is my first language.

Politically, at the time, women were getting into the New York City Police Department. When it comes down to it, I would have to say that I was just in the right place at the right time. I got into a CETA-funded pilot project and got training in air conditioning and refrigeration. That gave me some confidence. In 1978, many of the unions were opening up their apprenticeship for the first time in three years. Economically at that time here in New York, construction was like *pfhhht,* the pits, everybody had been forced to basically be without work. So it was a period when they were just starting to come out of that, and they were going to start taking apprentices.

In the women's community this woman Mary Garvin was talking about how journeymen electricians were getting paid $15 an hour. Many of the women that I had gone to school with—with their bachelor's degrees and with their master's degrees—were working for the Welfare office making $10,000 a year, or being librarians and needing master's degrees and Ph.D's. So their education is costing them a lot of money and they're not getting it back. Women's careers were like teacher, nurse, things of that nature, and I could see all of them were very low-paid. Looking at what they had to do and what they had to deal with, I eliminated a lot of those. Either because I didn't feel that I could do them, or I didn't feel comfortable in some way with them, or they just didn't pay well. When people in the women's movement were talking about women doing men's jobs and why shouldn't they be able to do it, it made sense to me. And since I had gone through that air conditioning/refrigeration thing, my level of confidence was there [that] I could do the work. So I felt I'd give it a try.

I applied for that apprenticeship program in June of '78.

— *Lorraine Bertosa* (born 1951) —

I lived in a real community-oriented neighborhood in Cleveland and was involved in community stuff and the food co-op. It was the Carter era. That whole couple of years there was a big drive to get women into the trades. I remember 500 women came down to this thing downtown at the community college. The unions must have helped sponsor this. That's how I found out when the test was and showed up and did all the paperwork.

I had worked with ex-mental patients for about five years. I was a cook and then a trainer. I'd been a waitress and all that. My boyfriend at the time was a roofer. He took me up on a roof job and let me help him.

Two things I liked that day. I remember sitting on the roof thinking, What an incredible view, from the top of the roof. And you could see what you had done. We could go to that house today, it's just down the street. And the roof's still on there.

Working as a cook or working with people, it's real hard to tell what you did that day. Or last week. Carpentry is concrete—it's so tangible! There's something good about that. I think a lot of male-type jobs have that piece in there. You have this thing you can touch and see and experience. Women have it in a smaller way in sewing and cooking, but they last so much less. *Pfffhh*, they're gone. People really remember these meals but—boom, it's gone the next day. "Oh, Lorraine, you're a great cook." You got a memory, but you don't have anything you can touch.

The year before I was exposed to the fact I could take this test, I had a job driving for a bathroom-and-kitchen-remodeler. I was doing only the delivery, and I kept trying to sneak in and do some of the work. The guys kept saying, "You don't want to do this, Lorraine. You don't want to do this." And I kept thinking, I'd like to learn how to put that sink in. I'd like to learn how to do this and that. Once in a while they'd let me help them, if they were by themselves and they needed some help.

So I started thinking. I didn't enjoy cooking as a living, you couldn't make the money there. And emotionally I found that I couldn't work with the people anymore. Growing up with alcoholism in my family, I didn't realize at the time that I was just continuing with this caretaking role on in my adult life. I couldn't figure out why I was shriveling up inside.

'Cause I hadn't addressed *myself* yet. What would you like, Lorraine? What would you like to do? Spring of '79, I took the pre-apprenticeship test and got into the union in the summertime. It was the money, yeah, but from Jump I was looking for something I liked to do.

— *Deb Williams* (born 1959) —

I came out of a trade school. They had started busing and they were integrating the schools. They opened the Boys Trade School to girls. Boston Trade High. I think that was probably '74, '73. I came the following year. The City of Boston sent out these pamphlets saying, What school you would like to go to: first choice, second choice, third choice. My first choice was the trade school, basically I think because a friend of mine was there. I was sick of high school anyway—math, English, basic stuff. I wanted to go for auto mechanics. I figured if I ever had a car I could tune it up.

When I got there, they told me that I couldn't take auto mechanics. It was the first day of school. I was in the cafeteria when the auto teacher says, "I'm not letting you in my class." And I said to him—I put my finger up and pushed him up against the wall—and said, "Look, that's discrimination, you can't do that." I don't know what the heck I was talking about. It was just a word that I knew. I think I was only 15 then.

He grabbed me and said, "I want to talk to you away from everybody else." He said, "It's in a cellar—and you're the only girl. There are all black boys down there." And he didn't trust them. That's exactly what he said to me.

So he scared me. He said, "Why don't you go in the paint department where all the girls are?" And I was like, "Well, all right." There were three girls in the paint department then. There was only five girls in the whole school.

Years later, when I thought about it, it was like, *You idiot!* Why should I have been so intimidated? I live in a project with all black people. It was not that they were black, I think, that frightened me, it was just all boys in general. Because he said it.

When I went into the trade school I was just a young kid, 15, 16 years old, and, like any teenager, this was party hearty. It was fun! It was better than sitting there and reading an English book, because we had half-day classes and half-day shop, which was great.

I goofed off. There was all boys in the school, what did you want me to do? I hung around with all the boys. I went to the plumbers shop, the carpenters, the electricians. I didn't take it very seriously. I did the shop work and stuff, but it never came across to me as a career, like the teachers in the school say, This could be a trade for you or a life for you to raise your family with. It was never brought across like that to me. I think the boys knew why they were there. The plumbers knew exactly they were going to be plumbers.

They had an assembly and they grouped you off, the plumbers, the painters, the electricians. The apprentice coordinators came in and all the apprentice teachers were there saying, Join a union. I was like, I don't even know how to paint. I knew when I graduated high school I had no intentions of going in the Painters Union and being a painter.

I was working as a nurses' aide in a nursing home. I used to go from school to a full-time job, 3 to 11. I was taking home $90 a week—I have check stubs—for forty hours' worth of work. And I was breaking my ass, it was a tough job. A lot of bedridden patients, you have to lift them. They taught you all that, but you don't get too much time to sit down. Back then we even did the bandages. You have changes, you have an hour supper, and then you've got to do it all over again. When I got out of high school I just went on with my life.

I had already been out of high school for a month. My sister Colleen graduated from the fifth grade and we went to her graduation. One of the teachers that she had there, I had known as a carpentry teacher from Trade. When I went to her graduation I went to see him, and he said to me, "Deb, did you go sign up for the apprenticeship program for the Painters?" And I said, "No, I don't even know how to paint, I can't be bothered."

He said, "Deb, they teach you." Now this is the only time I ever really paid attention to what he said to me. They may have said it in the assembly, I can't honestly say. I don't think they ever said it to me. But he put it to me, "Deb, it's all on-the-job training, you don't have to know how to paint. That's what it's all about, they teach you. You go to school and they teach you how to paint. They send you on a job and someone there teaches you how to do it."

And I was like, Really? He said, "If you don't go down to see Johnny Crisostamo today (he was the apprentice coordinator for the Painters) I'm going to break both of your arms and both of your legs." He gave me the address.

Graduation was over at 10:30 in the morning and I said, "Well, geez, I've got a skirt on, I'm all dressed up. Why don't I go?" So I got on the train and I went to Mass. Ave. and I went to see Johnny Crisostamo—and I couldn't even say his name, it took me years to pronounce his name.

— *Gloria Flowers* —

I never was very mechanically inclined as a young lady growing up. I didn't develop an interest in that until I more or less became aware of the economics involved in getting things fixed around the house. That's what more or less piqued my interest, when I was living at home with my mom. In high school, my courses were basically geared towards the direction that they would steer most girls—clerical, learning how to type, home ec. But I left home pretty early. I graduated when I was 17, in June of 1973, and moved out in September.

I attended Cleveland State University for one year, and got tired of going to college simply because I got tired of being broke and hungry. They'd post jobs that are available on bulletin boards and I got a summer job working at the Federal Building downtown Cleveland. And, boy! that little bit of money really did something for me, because after that I wasn't interested in going back to college. I just wanted to keep working and supporting myself, because college didn't seem to be holding my attention. I really didn't have in my family anybody to really push me, like I'm trying to push my nieces and nephews telling them, You can get through these hard periods. I know you're eating a lot of peanut butter and jelly sandwiches and hot dogs and going without, but stick it out—that kind of thing. I didn't get a lot of that. When I came home and told my mom I was tired she said, Okay. I started working and that was it.

I went from being a file clerk, to a clerk/typist, to a legal secretary, and I got to realize I was in a dead-end job. I wanted to go back to college, but I was strung out with a lot of bills—had bought a new car, you know. I said to myself, No way am I going to be able to quit and go back to school full-time and finish it in good time and, you know, still be a young person and enjoy myself.

It was 1979. At the time, Carter was President and they were really trying to push getting females into the local unions. They were doing a lot of advertising on TV and in the *Call and Post*. A friend of mine was going over to apply for an apprenticeship, and she says, "Come, go with me." Just a girl who worked in the same building as I did—you know how you meet people on the elevator and you talk and stuff? So it sounded interesting, and I went and I put in an application for the Electrical—the electrical IBEW's Local 38 was open first to take applications. But I got to thinking to myself. I says, Nah, I'm not going to apply for that one, because I'm a lot more scared of electricity than I am of water. So I waited for the Plumbers, which was opening up the next week.

My dad's a blue-collar worker. He's always worked in a mill or something. My mom was a domestic, our early years growing up. I didn't know anybody in plumbing. I wasn't related to anybody. I didn't have any friends or any relatives in any other trade either.

I believe ten of us initially applied. We were the first group.

— *Irene Soloway* (born 1954) —

My father and grandfather and uncle were small contractors—they weren't carpenters, they were contractors—for a period of time after the war. Really until the early '70s. It wasn't like a big outfit. They built two-family homes in Brooklyn and Queens, one house at a time. Then they'd sell it and build another one. They did all the troubleshooting and they hired all Jewish, all union out of the Brownsville cafeteria in Brooklyn.

I wasn't really exposed to too much of it, since I was the princess, you know. But my father used to take me to the jobs when they were in the finish phase, when the new floors were laid. So it was interesting, but I didn't grow up feeling competent with tools. In fact, quite the opposite, you know. I wasn't supposed to touch.

I went to art school in 1972 to 1976 in St. Louis and when I finished, I was working as a cocktail waitress for a year. In the bar where I worked, a whole bunch of roofers used to come in every afternoon and—literally—I was joking around with them and they gave me a roofing hammer. I never roofed with those guys. I told them I wanted to. You know—*I would rather be a roofer than work at this job anymore.* They laughed at me, but they gave me the roofing hammer and then, shortly after that, I started doing roofing with an all-women company.

It was roofing and painting. I didn't have any particular training. It was one of those all-women companies where everybody was improvising. There were a few people who knew. Like, you wonder now why anyone hired us at all. I guess we got up on the roofs in our bathing suits or something—maybe that was how we got jobs! But anyway, I just didn't make a conscious decision.

From there I did a lot of odd jobs, picking up here and there some skills, and then I moved back to New York and I was stripping furniture for an antique store. And then CETA was around, and I was in a six-month training in the boat-building industry—which was really interesting. From there, I worked in a boatyard on City Island. I was a laborer basically, cutting up scrap metal, and then eventually we worked on racing boats. They built boats for the America's Cup races and it was like a $4 an hour job. But I got to plane the keel, the lead keel of this racing boat. I worked there for a year, with mostly Portuguese men that were imported to work in the boatyards. When I was working in the boatyard I had some, you know, rape threats on the job that I had forgotten about for years, somebody else had to remind me.

Anyway, then I got into the union apprenticeship program.

— *Gay Wilkinson* (born 1940) —

Things were getting very bad economically here in Massachusetts in the '70s. We did have four children and we had a house. And the bottom just fell out.

I started looking around for a job in a very traditional way, working in a store or going into an office. All of these entailed spending money to make very little money. After being home as a wife and mother, I didn't have the wardrobe. Minimum wage was like $1.25 an hour and the competition for jobs was very keen, so some places were only offering 75¢ an hour. I called the Veterans Administration to see about hiring opportunities on the basis of the fact that I was a veteran. The only training program they had was at the Fore River Shipyard in Quincy. They were hiring people to be trained as mechanics to build their L & G tankers. This was '72, and they had an Equal Opportunity Employment program out there, but they hadn't hired any women.

I went down and I filled out the application. They offered $3.25 an hour. You would go through training, be certified, and then your pay would go up until you became a mechanic and get $5.35 an hour. Which was very good money. He explained that they certainly were an equal opportunity employer, but they wanted me to be aware that there were a whole series of things that I just might not like about the job. It would be all men, very dangerous, very dirty, you'll have to be out in the weather, the toolboxes are very heavy—I told him I would take the job.

I was the first woman hired there and I went into a welding training program. What was fortunate is, the instructor had worked with women during World War II and felt that women were an asset as welders, that they were much more patient, much more skillful in the long run than most men. So he was quite supportive and he kind of put a little protection-type thing there, where people would *think* to play with my machine or do things that would make me look bad. The training was for ten weeks. I certified in six.

I went to work out in the field, then came back into a training situation to further my skills. When I came in this time, there were six women that had been hired, all from a WIN program, which was a welfare program through the state. I was in my thirties and a good portion of these women were in their thirties. We had two young women, one was 19 and the other was 22. They put us in one little group, a six-person crew. Other crews went twenty, thirty people.

But an indication of the way it was going to go came the day that we went into our assigned area and we were given these wooden toolboxes. We were told to pick them up and bring them to the lay-down area where we were going to be working. We had a very young male foreman, and we got about half of the way over and one of the young women was really having difficulty carrying her toolbox. She put it down and sat on it and he very nicely picked it up and put it on his shoulder.

At that point, I told the other women, "Put your toolboxes down and sit on them." We did. He came back and said, "Excuse me, but get those toolboxes." And I explained to him, "If you're carrying her toolbox, you will carry all of the toolboxes." And we sat there for the rest of the day on those toolboxes, until he picked up every single one of them.

I think it was almost an instinctive thing. I mean, the men there thought this was a joke. There were whistles, catcalls, everybody's waving. I just decided that, as far as women go, or my particular crew—I can't say I really thought women on the whole—if we didn't set some guidelines real quick, this thing was going to become a joke. And it was very important to the women who needed the job—I needed the job myself.

They asked us to work under some very, very unsafe conditions. We had lines that were frayed. We had days that it was just so cold and so wet that your feet would actually turn white, and yet you were expected to stand there and weld for hours upon hours. In the time that I was there, I saw seven men killed, and they were killed for reasons that were totally ludicrous.

While I was working in Fore River, my husband had gone to work out in Michigan, and Michigan was a strong union state. He called me because they needed certified welders for the Boilermakers and with my certifications, he certainly felt that I could go to work. This would mean going from $5.35 an hour to $21 an hour with union benefits. I arranged for a sitter for my children and I drove to Detroit.

I went out to the Boilermakers hall, walked in, and I said, "I understand you're hiring welders." They said, "Yes," real excited, "who's this for, your husband?" I said, "No, I'm the certified welder." Immediately everything got very quiet. This very angry man came out and said, "I have to hire you because we've told you that there's a job available, but you won't last three goddamn days, I guarantee it."

The next morning I reported at a power plant 140 miles north, to the steward on the job who already knew I was coming. The first man that we came to, told this man to get his paycheck, he would not work with a woman. Put his tools down and that was it. This happened with two other people. Finally there was this young fellow who said, "I can't quit, I need the money. I'll work with her, but I will not talk to her, I will not eat lunch with her, have my coffee break," blah, blah—listed the whole thing. "She can work," he said, "but we will not be partners."

For three weeks no one except the general foreman spoke to me. If I sat down, everybody would move to another area. If I went up to work, they would actually try and move their work. It was getting very discouraging, very depressing. I was working six days a week, twelve hours a day. My husband had gone home to be with the children, so my only contact with family was by telephone. It was very lonesome and there were no other women. I mean, *no other women.*

At the end of three weeks, we had this one particular job where you had to go between the thirteenth floor and the twelfth floor and weld what they call a "buck stay" to the boiler. You had to be able to do it left-handed and holding a shield. The man who was supposed to be my partner couldn't get his arm in there between the floors. So I told him, "I'm going to put the safety line on and you're going to have to hold it."

He looked at me very strangely and he said, "You're going to let me lower you between the floors?" I said, "Whether or not I live or die is going to be totally up to you. You are my partner, this work is our responsibility, and I'm the one that can fit in there. So I'm going to hand you this end of the rope," I said, "and I am going to more or less take on faith the fact that you will not let me go."

Between the grating and the boiler, it is just open all the way down to the bottom. Thirteen floors. Straight down. I took my shield in one hand, I put my glove on, I got the welding rod ready to go, and I said to him, "Lower me down." And it was like this very pregnant pause. I went down. And I'm here. So we know he held onto that rope.

When I came up and undid the harness, he asked me if I would like to go for coffee. It was the first human contact with any kind of civility to it at all. I acted like the three weeks just hadn't existed, went for coffee. He sat there, asked me why the hell I would ever consider doing anything like this. I told him I had children, my husband hadn't been able to get a job, I had the expertise and I felt this was the way to use it. Then he talked to other people, and what evolved was a very good relationship with most of the men. It was just mind-boggling to them that women were working in the shipyard on the East Coast. And I was very surprised to find out that, with all the work going on out there, women just hadn't made any breakthroughs at all.

I worked out in Michigan for almost three and a half years. Most of the time I didn't have to work more than five months and I could be home for seven, because the wages were just absolutely unbelievable. I would work seven 12's, clearing $2,200 after taxes and expenses. While I was home those months, Bill would go out to Detroit and work for the Ironworkers. Meantime, I'm getting all this experience and I'm coming back to New England and—'76, '77—there are jobs *here* with the Boilermakers.

I couldn't even get in the hall. They wouldn't even let me in the door—literally. So I'd go back out to Michigan. I was traveling 900 miles one way to work.

PIONEERING

T HE CIVIL RIGHTS ACT OF 1964 prohibited discrimination in employment. The next year President Johnson also signed Executive Order 11246, prohibiting discrimination in government contracting and requiring affirmative action for minorities on federally assisted construction projects. It established a monitoring agency in the U.S. Department of Labor, the Office of Federal Contract Compliance (which later became the Office of Federal Contract Compliance Programs). In 1967, EO 11246 was amended by Executive Order 11375 to include women. But OFCC[P] set specific goals only for the hiring of minorities, not for women.

In 1976, a consortium of women's groups from across the country sued the U.S. Department of Labor for failing to provide the required affirmative action. The ultimate result was that on April 7, 1978, President Carter issued affirmative action regulations (41 CFR 60) that expanded Executive Order 11246 to cover "all construction contractors and sub-contractors who hold a Federal or federally assisted construction contract or subcontract in excess of $10,000." The regulation set "initial" goals and timetables "intended to provide immediate equal employment opportunity for women in the construction industry." They started at 3.1 percent and escalated by 1981 to 6.9 percent. These were never quotas or preferences, which would have been illegal.

Explanations made clear that the jobsite goals would increase as more "women journeypersons and advanced apprentices" were available. These goals applied to the contractor's entire workforce, including workers not on the federally funded site, and established that the "hours of minority and female employment and training must be substantially uniform throughout the length of the contract, and in each trade." A committee was to recommend more permanent goals and timetables. Further provisions required contractors to

- "ensure and maintain a working environment free of harassment, intimidation, and coercion at all sites, and in all facilities at which the Contractor's employees are assigned to work."
- "where possible assign two or more women to each construction project."
- provide nonsegregated facilities except where "separate or single-user toilet and necessary changing facilities shall be provided to assure privacy between the sexes."
- hire among racial groups relative to the racial make-up of the population.

On May 12, 1978, the Department of Labor additionally set out goals and timetables for apprenticeships (29 CFR-30), establishing that each new class of apprentices should include females "at a rate which is not less than 50 percent of the proportion women are of the workforce"—or roughly 20–25 percent of each class. These regulations were also extremely detailed about procedures to be followed to ensure equity.

The regulations were the result of aggressive pressure from the women's movement. The Department of Labor, involved since 1971 in outreach efforts to bring women into "nontraditional" jobs, concluded that voluntary affirmative action plans had not been successful, whereas those with specific goals and timetables had. Local goals for hiring women were already in place in Seattle, Madison, Wisconsin, and California.

By 1983 women were 1.8 percent of the construction workforce.

Doors, Windows, Locks

I felt like we were, you know, truly pioneering. I went shaping around with a black woman named Martha Clanton and we wore hardhats to shape and we had a brick thrown at us. But mostly it was very confusing and a lot of run-around. You go to the fourth floor, to the foreman—now he's on the seventh floor. One company, the two supervisors just rolled on the floor with tears in their eyes—they thought it was the funniest thing in the world.

There was a lot of resistance. I didn't feel that we would ever get a job this way. It did seem quite impossible.

—*Irene Soloway, New York City*

In 1977, after completing a federally funded, four-month, hands-on training program in construction skills—Non-Traditional Occupations for Women—I called up the electricians union ready to become an apprentice. I expected the procedure to be simple. The man on the phone—having determined that I was white and not covered by any federal requirements at the time—laid it out straight, "The unions don't want you, the contractors don't want you. We'll call you if we need you."

So I signed up, with another woman who'd graduated from the same CETA program, for a night class in electricity at the local trade school (the fee was $1 per semester), and called electrical contractors listed in the Yellow Pages, trying to get on as a helper. When that didn't work, I housepainted until a friend-of-a-friend hooked me up with a non-union electrician writing houses. On a Friday afternoon in April 1978, I got a phone call from the apprenticeship office: "Be here Monday for an interview."

—Susan

By setting percentage goals for women to be brought into jobs and apprenticeships in the construction industry, the 1978 federal affirmative action regulations contradicted the commonly held cultural assumption that women were unqualified to do that work. Recognizing that doors had been unfairly locked, they created a key. But the existence of a key did not by itself open doors.

In 1978, the unionized sector of the construction industry offered higher wages, safer conditions, and better benefits packages than the non-union sector. In the skilled trades, unions were also highly regarded for the quality of their training programs, funded by members' dues and contributions of signatory contractors. It was these union jobs and apprenticeship slots that were targeted by efforts to bring women into the industry.

Each trade—carpenters, electricians, plumbers, pipefitters, ironworkers, sprinkle-fitters, painters, elevator constructors, operating engineers, sheetmetal workers—had its own international union organized geographically into districts and locals. The unions' power came from their political leverage to ensure that construction projects, particularly in large urban areas, would be built by union labor, and by their ability to control access to those jobs through their hiring halls.

There were two doors to union membership. An already skilled tradesperson could "buy their book," that is, pay an assessment to the union and enter at journeylevel. Or one could be accepted into an apprenticeship program jointly run by the union local and signatory contractors and graduate as a journeylevel mechanic, upon completion of three to five years of prescribed training. Most women sought access as apprentices.

Entrance procedures and requirements (such as age limits) varied by trade, region, and local. The process to become an indentured apprentice often included submitting an application, taking a test, having an interview, and finding a first job. Any of these could be points of derailment. Some of the training and advocacy programs established to assist minorities—who had gained similar access in the early 1970s—expanded their outreach to women; and new organizations emerged, immediately prior to and following Carter's Executive Order, designed specifically to address women's particular issues of access.

From ads on the subway, MaryAnn Cloherty learned about a nine-month full-time training program to prepare women to enter union apprenticeships. After calling the number she had written down, she was assigned to the program in Plymouth, Massachusetts, just south of Boston, in early spring of 1978.

I REMEMBER THE DAY I DROVE down. It was very exciting, because there were thirty other women there and they were all there for the same reason I was—they wanted to get a skilled trade. It was the possibility of earning a living, to have some decent wages, to have a skill, to feel as though I didn't have to depend upon the state, welfare, to support myself and my kids. Being on welfare is $89 a week, I recall—that included my food stamps. It was just a horrendous situation. I saw going into this training program as a way to become independent.

We did mathematics. We did language of the trades. We did role playing. Five days a week. We had four teachers, each of whom was a union tradesperson. Of course they were all men. But we did have a counselor who ran the program. She was an educator, Kathy. She was a black woman from the Plymouth area. Terrific woman.

That training program was very positive. We went out and—we had a gymnastic instructor—we ran the tracks, the old railroad tracks in Plymouth. We were running five miles every other day. We were lifting weights. We were really empowered. We were ready to take the world on! They gave us four choices. I was more interested in carpentry.

We went down to a battered women's shelter. It was an old farmhouse and we got to go in there and put in new pulleys and ropes in the windows and fix busted doors. It seemed very exciting to me to go into this old ramshackle farmhouse and spend a couple of weeks there with a group of women making things work. Simple things, but the overall effect was to take something that was derelict and to turn it around, into something useful and helpful. Then there was the whole dimension that it was a battered women's shelter in a secret location, and there were these women and kids there.

There were thirty of us that graduated. It was subsidized, it was one-third federal funds, one-third state funds, and one-third union funds. I was told by the director of the program that they each contributed $6,000 per trainee. So that would be $18,000 per trainee. Times thirty.

For Lorraine Bertosa, who wanted to become a carpenter in Cleveland, the process worked smoothly. At a heavily attended public meeting aimed at recruiting women, the unions announced the dates for application and testing. The carpenters union ran their own hands-on, exposure-to-the-trades training session, not just for women, but for all new apprentices.

S PRING OF '79 I TOOK the pre-apprenticeship test and got into the union in the summertime. I ended up 13 on a list of 500 people. It was a bizarre, stupid test—it had nothing to do with carpentry. I got a job right after that pretty much.

We had a four-week pre-apprenticeship class where each week they took us to a jobsite. We built a toolbox that I still have, and a sawhorse. You learned the basics. We had class, we saw the jobsites. We went a five-day week, thirty hours or something—got forty bucks a week to do that. It really gave you a flavor for the tools a little bit, for some of us. I thought the jobsites were just a gas! They sifted through a few people during those four weeks, which was smart for them.

They were really pushing women. We had five women out of twenty students in my class, which was incredible. The first woman in the local, in the union here, was a black woman. She was just one year ahead of us.

At roughly the same time Cheryl Camp entered the electricians union in Cleveland. As an African American and the only woman in her class, she was a "trail blazer." Cheryl worked a variety of jobs after high school—receptionist, secretary, switchboard operator—and became interested in computer repair. She had been taking electronics courses part-time for a year and a half and was working at Allen-Bradley. Although she had no family in construction unions, it was an uncle who alerted her to the apprenticeship opportunity.

M Y UNCLE CALLED AND SAID that they were taking applications and to go down and put an application in. Because one of his friends who's an electrician told him about it. I was 23, so I got in right under the wire. In June they took the applications. The test and interview was in '78.

The entrance test, it was a test of agility and comprehension. We had a briefcase-looking device, and it had pegs in it, and to test your agility we had to flip the pegs over in a hole. We had to move the pegs as quickly as we could from one section to the next section. What was really weird was—all these guys were in the room and I was the only female when I went to test. They were all looking at me like, What are you doing here? I beat them all on the testing because I have small fingers. And because I was doing switchboard operating and it was the old type, with the plugs. I had to work with both hands to do that, and that was basically the same thing. So I was working both hands, working those pegs, and they were amazed. The guy next to me—I know he had to fail that part of the test, because he was watching me instead of taking his test. I did pretty well, and then went to the interview.

The interview was a panel of people. It was the business, agent, the president, two contractors from NECA [National Electrical Contractors Association], and they were all interviewing me at the same time, asking me these questions. There was this older guy, he was a contractor, he's got to be dead now, because he looked like he was on the verge then. He kept asking questions geared towards discouraging me. He said, "Are you sure this is something that you want to do? This is going to be really cold conditions. Have you ever lifted a piece of four-inch rigid pipe?" And I said, "Well, I don't know. What is four-inch rigid pipe? I don't know how heavy it is. I don't know if I could lift it or not."

"Well, how about heights, are you afraid of heights?" I said, "No, I've done housepainting before, so I'm not afraid of heights." Everything he said was to discourage me. And that's when the business agent intervened and said,

"Look, we'll just ask regular questions." And after that the interview kind of narrowed out, so it became much more pleasant. But I think he was one of my motivating goals to make me decide that I'm going to do this, just to show them I *can* do this. I was supposed to start in that class of either September of '78 or February of '79, but there was a wait. I started working in July of '79.

Although across the country many of the first women who graduated from apprenticeships entered the industry in 1978, there were some union locals where in anticipation of affirmative action, women were brought in slightly earlier. In others, because of weakness in enforcement or the economy, the first women entered somewhat later. Paulette Jourdan and the other woman who entered the plumbing apprenticeship with her doubled the female membership of their Bay Area local. But first she had to pass an exam that tested "basic skill levels in English, math, and spatial relations."

I HAD BEEN OUT OF school for twenty years. I must have sat up every night, six hours a night, I guess, for two months, just to get that math back. It was algebra up to geometry. I got enough of its back so I did well on the exam. And thanks to affirmative action I was only competing with the people in my own ethnic group, which helped. There are guys there, it's part of their heritage to have the advantage of having someone that went ahead of them. There was one white woman who scored at the very top of the white guys. I don't know how she did that, 'cause I heard that some of them had the test. I really admired her. Out of a possible 200 I scored around 150, which I didn't feel too bad about. I was at the top of my group. There were two slots open for blacks and I got one of them.

It was all very clear. There were twelve slots, six for whites and two for the other three minorities that they allowed. Which is black, Hispanic, and Asian. I think they had one Native American—maybe they created a thirteenth slot. I was feeling lucky because I remember a time when it didn't exist, there were no slots for anybody else, so I was glad to have a chance. That's basically how I got my foot in the door.

There was no oral interview at that time. They accepted me because I passed. They wouldn't let me start school because there were no job openings at the time. I waited about ten months before I actually started school and started on a job. It was at the beginning of '82 and I started work at the end of '82, December.

Apprenticeship programs were not generally open for applications year-round. The narrow window of time when applications could be made—such as a two-week period every two years—had been one way for unions to limit access. It was necessary to know when to apply. Members, who might want to bring in a son or a nephew, were generally informed well in advance. Unemployment offices and advocacy groups tried to play a similar role by bringing that information to interested women and minorities. In New York City, Cynthia Long was assisted by a journeylevel carpenter who'd recently moved from the Bay area and who was organizing a pilot project, Women in Apprenticeship, modeled after a similar group in San Francisco. Cynthia applied to the electricians union in June 1978, when a long construction bust was ending. The new apprentices would begin work that August.

IN TERMS OF THE APPLICATION process, Mary kind of prepared us for the fact that it would be hard. She would say, "As soon as they tell us the line is starting, we're going to go out there and camp out on the streets." Having come from middle-class America I'm saying, Camp out on the street? I don't even camp out in the woods, why would anyone camp out on the street? I just had to kind of put that aside and say, Okay, if this is what this person feels like we need to do—okay, I'll do it.

We went out Wednesday night, and the applications were being given out Monday morning. What is that—Wednesday, Thursday, Friday, Saturday, Sunday—five nights. As soon as the line started, the union had to do something to maintain order. What they did is, you had to sign a book and you had to show your ID to verify who you were, and they gave you a number. Then they could come at any time and say, Where's number 56? If you weren't there, you had like five minutes to get back. So what would happen is, if you had decided to go home to Brooklyn to take a shower, it wasn't a good idea, because you wouldn't be able to get back. They told us if we weren't there and we didn't report to them, that they would cross our name off and that we would have to go to the end of the line.

We literally stayed on the sidewalk. We had this whole system. Obviously there weren't sanitation facilities—so this was my early exposure to not having sanitation facilities on the job. What we did is, we would go into like your diner-type place, order some food to go, and use the bathroom. And just kind of let personal hygiene lapse for those five nights. Part of the reason we got on this line was because we were told that they were only going to give out x number of applications.

I think that there were eight of us. Mary Garvin was there to kind of keep us together. And she also prepared us for the fact that there might be physical violence. She was talking about passive resistance and anti-war—you know, coming from that kind of background, she was telling us different tactics. And I was like, Wow. It made my mind spin. I was saying, If this is what she thinks might happen, okay, I'm going to listen up. But I couldn't conceive it, I just couldn't conceive of it.

As it turned out, it was relatively calm. Because once the minority men got on line and the women got on line and we wanted to be together—so that we wouldn't be just one woman harassed by all these guys—once we did that, then the sons of the electricians had to get out there and camp on line, too. They made it like a big party. I remember some of the boys getting pretty wild.

And I remember, too, the families of electricians. Because this area is called Electchester. The housing is built with Electricians funds and a lot of electrician families live out there because it's a co-op and the rent is very low, so it's making affordable housing for working-class families which was really, I thought, a very good thing to do. I remember some families coming by and asking us, "Why do you want to be electricians?" Mary had kind of prepped us for this and we each gave individual answers, 'cause all of us had different motivations for doing this.

Not yet knowing Cynthia or many others on that sidewalk, Melinda Hernandez joined the line with another group, All Craft, a two-month CETA-funded training program "for minority women and welfare mothers."

MOST OF THE WOMEN THAT were at All Craft, we all slept in the same corner, and Cynthia was with a group of women—they were up the block, we were down the block. We'd visit each other in the course of the day. It was like 90 degrees out, thank God for that. I think Cynthia was one of the first in line. There was like a thousand applications given out and she was number 32 or something. And I was down at the middle, 362 was my number.

The men were just partying and drinking and you know, smoking pot and everything. They were getting blasted. To them it was a big party. To us, we were not about to get blasted or drink or anything because we had to know what was going on around us.

They did things to discourage us. Like they pissed on some of the women's sleeping bags, and they kind of jumped the curb that I was sleeping on with a car. You know, just to scare us. I mean here you are sleeping and a car jumps on the curb. They tried to get us off the line, but we went through with it because—then we were mad. It was like, first we wanted to do it because we wanted to try, you know, for an opportunity. But then when we seen that there was resistance simply because we were women, we took a stand. We weren't going to be scared off. I believe there were like thirty women on that line, but there were only four or five women taken into the program at that time.

You were afraid because you didn't know these people. You didn't know the women, but you felt comfortable because they were there for the same reason as you. But the men you didn't know and you didn't know what they were capable of. I've never been exposed to such direct hatred like that. So, you know, it was an experience.

I remember, it was my twenty-second birthday on the day I received the application. And there was an age limit, 18 to 22, so I just made it. This woman that I knew who was in All Craft too, she had applied and been accepted. But her husband beat her up because he refused to let his woman work in construction—and she wanted to. She came into All Craft and she said that she couldn't accept the job with the union. I was really furious. I mean, I had never really experienced outright oppression of females. My father was a very mellow kind of guy, you know. He might have been macho in his own way, because he's Puerto Rican, but he wasn't macho outright. As far as men beating up on their wives I wasn't aware of that. She was a Puerto Rican woman, too, and I said, Well, one of us is going to make it. You know, he stopped her but nobody is stopping me.

Many joint apprenticeship boards used interviews to rank those who passed the qualifying test. Although interviews allowed the applicants to be seen as individuals and not just test scores, they were also highly subjective, particularly when handled by only one person. Candidates could be disqualified, or placed so low on the list that they would never get a spot. Maura Russell found being an articulate college graduate strongly in her favor in being selected over other women to become an apprentice plumber. As one of the licensed trades, becoming a journeylevel plumber—whether union or non-union—included the additional requirement of passing a state exam after working a required number of hours under the supervision of a plumber holding a master's license.

I HEARD THAT THE UNIONS WERE being forced to let women in. At that time I thought, Well, a licensed trade would be a more useful trade to get involved with. I mean, having studied a fair amount of women's history—thinking that this might be a small window of opportunity for women to get into the trades. Then it was down to plumbing and electrical, and electrical did not attract me. It was mostly a choice just to get a license. At the time very clearly thinking that, if the door ever got shut again, it could *really* be shut. Unlike carpentry or bricklaying or painting, where some women can go and choose to just do that and teach themselves how to do it and then be able to work at it, you can't do that with a licensed trade. If you don't go through some very specific apprenticeship with a specific person, you're shut out. If there weren't any women with the license, that would require finding some male mentors—which is not as though there aren't any around—but it's a more dependent thing.

It wasn't like I got some kind of tremendously welcoming response from the plumbing union that decided me to go in. I mean, the guy was paternal. I very much felt in the interview that there would be a class bias in who he would take. It was really clear in the way he talked about women he had rejected, the way they spoke, that they were not *ladies*. He actually referred to one woman in the electrical union in trying to tell me how I should be if I was on the job, "*Oh my God, the way she talked! The mouth on her!*" He said, "She doesn't act like a lady and she won't be treated like one."

He clearly, clearly rejected women who didn't fit his image, if somebody was just rougher in speech and swearing or somehow coming on stronger. I remember being struck by that at the time, thinking he was really trying to pick people who he sees as polite and compliant. Not consciously, but that's what he's attracted to. It was very much—he was being compelled to fill a few slots. He didn't think women should be in construction, he would never want his daughter to be in construction, it was just not a place for any "lady" to be. He said that very specifically. He was giving me helpful hints—I should wear gloves, so I don't ruin my hands when I work.

Making it to the interview was not always easy. Nancy Mason, a schoolteacher in Seattle looking for more secure employment, was guided by advice from an employee of the Washington State Employment Security Department who had visited her class to talk about career opportunities.

I DECIDED THAT I WOULD APPLY for both the plumbers apprenticeship and the electrical apprenticeship. I did that in 1978 and was warned by Alice Lockwood that both of these programs systematically destroyed people's application component. They'd either lose your transcript or you wouldn't get your doctor's note or you didn't get a high enough grade on the state aptitude test. Something was always incomplete in your application process so that you never got an interview. I sent everything to the training director registered mail and I would call him once a month—as Alice suggested that I do—to check to make sure that all my application stuff was there. At one point, he told me I had a problem with my math. I had college algebra, he had my transcripts. And he wasn't sure that my math was good enough to qualify. I said, "Well, what about my college algebra and statistics courses? This is way beyond what your requirements are." He finally agreed that maybe the math was okay. Having Alice tell me almost a year ahead of time that he systematically did this stuff, I was very much aware of what he was doing.

I knew the interviews were going to be in the first part of May. I had a male friend who had also applied. My friend got his interview letter, so I called up to ask the training director where my interview letter was. He just mumbled something about how my letter must have got lost in the mail. My interview was two days from the day that I called him, at eight o'clock in the morning. I told him, "Well, I'll be there." They only interviewed, at that point in time, once a year.

I showed up at 7:45. He must have forgotten that he talked to me and I told him I was coming, because I went to check in, and my name was already crossed out as a no-show—at 7:45, which was fifteen minutes before my interview.

I had my interview with a management and a labor trustee. I decided because of my name being crossed out that I might actually pursue them, because Alice had been looking for someone to sue them anyway. About a week and a half later, I got a letter from the apprenticeship program. I was astounded to see that I was indentured into the program—which ended any possibility of pursuing what happened during my application process.

Eighty people got in that year. There were nine of us [women] indentured the same year. We were the second or third year of women coming in, but the first year of any kind of serious numbers. That was a big deal. It caused a big ripple.

Admission to an apprenticeship program had different consequences depending on how apprentices in that particular trade got jobs. In programs where the apprenticeship director was responsible for assigning each apprentice to a contractor, admission meant employment. In other programs, where getting hired was the apprentice's responsibility, admission had less value. It meant only the *right* to look for a job for a given period of time, a "hunting license." Those with a father or uncle in the industry could of course turn to him for help in finding work with a contractor. Those without family connections needed perseverance and luck. Kathy Walsh became an apprentice carpenter in Kansas City. She learned of the opportunity through efforts of the Urban League, in the winter of '78. But once in, she was on her own.

I HAD THREE KIDS AND one of them a new baby. And a house payment. And car payment. And a husband that disappeared. I had always done office and clerical-type work before. I was desperately looking for some other way to make more money on my own and went to this seminar put on by the Urban League and met Lisa Diehl, who is one of our very first carpenters here in Kansas City, who was there speaking. The money sounded really, really good.

Lisa is a very small woman, maybe five foot, maybe 100 pounds soaking wet, and she was sitting there in this chair and she looked so—like a cute little girl. I'm six foot and I'm a big person, big woman. I thought, Well, if she can do this kind of stuff, I can do this kind of stuff. She was an apprentice, she was just starting, so she was very enthusiastic also.

I went and signed up with the Carpenters apprenticeship program. Lisa had the information on where to go and what to take with me. The apprentice

Third year apprentice plumber Marge Wood brings a 36″ wrench with her to the Urban League for her first public speech about working in the trades. Photo copyright 1980 by Tom Bernthal.

coordinator was very nice and took my papers and said, "Here's a list of our contractors. If you can get somebody to hire you, we'll accept you as an apprentice." Anybody could apply, but the trick was finding the job—and they didn't tell you that. They gave you a list.

I didn't even know enough at that point to even ask, how do I do this? or what do you do?—anything like that. So I went around to a hundred contractors' offices and put my name in. A couple had an application form but for the most part all they wanted was my name and phone number. I did all that. And nobody called me.

Doing clerical work at the time, Kathy drove to contractors' offices on her lunch hour, in her work clothes.

SKIRT, HOSE. I'M SURE THEY were impressed. Probably occasionally I had heels on. It never even occurred to me to go to a jobsite at that point. I did this off and on for about four months. I was going nuts doing what I was doing and trying to work. I had a one-year-old, a ten-year-old and an eleven-year-old. It was rough. I lost it. That was back in the good old days, when if women were having problems and not coping, your doctors gave you Valium. I went through that whole thing back during that period. Finally, at some point, I said, This is not going to work like this.

I started calling the apprentice coordinator two and three times a week. I started going to jobsites. There were only about four, maybe five, other women carpenters at the time and I had met maybe two or three of them. One of them said, "Well, you're never going to get a job going to their offices, you've got to go to the jobsites." So then, here I am, in my blouse and skirt and heels, and I'm trying to go to jobsites. 'Cause I had to dress up for work.

I remember one time going on a jobsite saying, "I want to be a carpenter, I want to do this kind of work, I just want you to give me a chance. I will do whatever you want that needs to be done." Which was not the right thing to say to this superintendent that I was talking to. He took me into his office and told me what he wanted me to do. If I would fool around with him, he'd give me a job. He didn't know whether it would be as a carpenter or not.

I learned *not* to say, "I'll do whatever you want me to do." I learned to be more selective with my words and say, "I want to be a carpenter." And I'd try and find out a little bit about what they were doing. I was learning at this point that you did sheetrock or you did metal studs or you did concrete—different areas. I'll never forget going to this one drywall contractor—this guy literally locked the trailer door behind us when the two of us went into his office for me to fill out an application. I was like, Oh my God! He never touched me or never specifically said he wanted to have sex or anything like that but he was—like in the movies what you see—you're backed up to the wall and they're standing there with their arm against the wall where you really can't go anywhere, right there in your face! I was so flustered. I don't remember how I got out.

In San Francisco Donna Levitt could look to Women in Apprenticeship, an advocacy group assisting women into the trades, for counseling and advice on landing a job. Still, the enforcement of affirmative action standards was key to her first success. She worked evenings as a bakery janitor and applied at construction sites during the day.

I GOT MY LETTER OF SUBSCRIPTION from the Carpenters Apprenticeship Program and I spent about three months going job to job. It was horrible. Find the foreman and tell him I was looking for my first job, that I knew how to use tools and would he give me a shot? Would he be willing to hire me? I remember at the time Moscone Center was being built and there were a lot of women being hired there. That man that was in charge of hiring seemed to get off humiliating me. I'm 5'1", which has also been a difficult thing. Because when you're judged on your appearance, it's easier to make the assumption that I couldn't handle the job or I had unrealistic expectations that I could make it in the trade.

Mostly they would say, check back in two weeks. I would usually check back in four or five days. I remember making lists of where I had been and what date and who had said what and what was worth following up on, and getting more leads from Women in Apprenticeship. Finally I got a phone call back from a woman who was an EEO officer for a large construction company, who said that I could come to work on a new high-rise that was being built.

Although many advocates and training programs played important roles in helping women gain access to the skilled trades, there were also, from early on, funded efforts that succeeded in creating the veneer of compliance more than its reality. A pool of applicants available and waiting did not change women's economic situation or the gender make-up of the construction industry. To make an impact, training programs had to come through with placements in these jobs with high earning potential. When MaryAnn Cloherty and the other women from her nine-month pre-apprenticeship program applied to the unions, they did not find the open door they had expected. MaryAnn approached one of the carpenters locals.

A HIRING AGENT SAID TO ME, point blank, "We had the coloreds forced down our throats in the '60s and we'll be damned if we have the chicks forced down our throats in the '70s." Direct quote. And no back-up from the [training] program. None whatsoever. Once you were out of that training program that was it. You were on your own. Good luck. Pat you on the back.

I filled out papers when I was there, but he was basically blowing a lot of smoke in my face. I still filled the application out.

She was notified to attend apprenticeship school, which she did. Unlike her male classmates, though, she did not have a carpentry job.

I HAD NO IDEA WHAT THE procedure was. Nobody said, Go out and find a job. Certainly not. I really got the sense that the only work that I would get would be work where there was some type of compliance going on and that they would need a woman—and that's when I would get a call. I did go to apprenticeship school at night, two nights a week, and I did eventually get a call telling me to go to a job one day. I suspect the only reason they sent me was because it was a federally funded job.

I had a party down at my house, I invited all the women from the training program. It was the summer of 1980. Out of all of us, only two of us were working in union fields, even though the unions had funded our training. Two out of thirty. A couple were picking up side work here and there, but it was like their boyfriend's business or something. Everybody else had gone back to secretarial work or waitressing or whatever. They hadn't gotten accepted into the unions. It was a big disappointment.

The signs were still on the subway, but also we knew that our director had resigned in protest because nothing was being done to place the women who were coming out of these programs. When that next group of women appeared, she protested the fact that the first thirty women who had just gone through this nine-month-long training program weren't placed.

I was in the first program. The following program they cut the training time down from nine months to six months. After that, they cut it down even further. The training program—in many ways it was seduction and abandonment. We were given all this wonderful training, given all this preparation to enter the trades, and then really not allowed to enter them. I didn't expect we'd be overrunning the places, but I expected to see women on jobsites. There was no follow-through.

Some early tradeswomen did have the advantage of family connections. When she was going through a divorce, Mary Michels' ironworker father suggested she learn welding and got her into a twenty-week school in Los Angeles taught by a fellow ironworker. Her father offered to use his connections, once she completed the training, to get her a job. Not in construction, but with a shop. "He did not feel that women belonged in the field, there just were no women out there." After completing the program and getting her certification papers, Mary ended up—with support from the men in the welding school—not in a shop situation but on the same construction job as her father, being paid journeylevel rate.

EVERYBODY THAT WENT THROUGH THE Pacoima Skill Center had gone to work at the brewery, Busch Gardens. They had a big project there making beer tanks, and most all the guys that went out of the school and graduated, they were all working there. They would come back and visit us and show us their checks and tell us to come on out there. I wasn't going to go because I think it's too heavy and then they kind of says, "Look, there's a lot of guys out there that aren't big either and *they* can cut it. Give it a try," and so I did.

I went to the union hall—and times were really good—and told them I wanted to be dispatched out to Busch Gardens. I was considered what they call a permit hand. I was not an apprentice. I was a certified welder and I was working at Busch Gardens as a permit hand.

Unable to reach her father by phone, Mary didn't run into him at work for days because they worked in different areas of the large site.

FINALLY, BY THE THIRD NIGHT, someone on the job told him I was out there, and they said they thought he was going to have a heart attack, he was so shocked. He came over and said hi to me, and that was it. I don't know if he ever approved or what. I got into the Ironworkers because basically I felt I had no choice. The opportunity was there to learn how to weld and I needed the money and there was no other way. What were my choices? I didn't want to live in poverty. I wasn't interested in welfare—I would have been eligible for welfare with what I earned, so that says enough about what I made. And I really was not interested in getting married to have someone support me, because I'd end up having to work anyway.

My dad, that was a connection. I never would have thought of it on my own. My dad was in it and he was a welder. There was so much construction in the summer of 1980, they could not get enough welders. You go in and ask for a job and it was yours. They even gave us presents. We had Bud points for going to work every day. Each week you'd get, like, 500 and then they'd double and triple and you could get all kinds of tents. Some of those glasses way up there are from Busch Gardens from years ago.

When the Busch Gardens job ended after a year and a half, Mary applied for membership in the Ironworkers local. A number of the men in her position tested in at journeylevel and bought their books with the union.

THERE WAS ONE NIGHT A month you could go down there and take a test to get in, but they wouldn't let me take the test, the people down at the union, for whatever reason. The union was very supportive of me, they definitely wanted me in the union, but they wanted me to go through the apprenticeship program. They gave me a hard time about it, but I actually appreciate that part, that I went through the apprenticeship program rather than buy my book. It's given me much better background. And there's nobody that can argue that I never went through the apprenticeship program. There's respect there, for the three long years that it took to go through it.

They needed to fill females. I don't know what kind of pressure it was. I call that Before the Reagan and Bush Era. Before those guys, there was a real change. And then when those guys came in, from then on—downhill. There was definitely pressure on the union. I was 35 at the time, but I was the child of an ironworker so they wanted me in.

Diana Suckiel asked cousins who were members of the plumbers local in Kansas City to sponsor her when she joined.

I'VE KNOWN THEM MY WHOLE life. They're good people, but they're very traditional. They see women in certain roles and that's where they should stay—and if that's the way they want to feel, that's fine. But the night that I was signed in, when they indenture you and you sign your contract, they were there. And they said, Yes, she's okay. They have to vouch for you.

In our class, there was probably a token person for every minority slot. We had three black women and two white. Then, as far as the men in our class, we had one black, one Indian, one Hispanic. I think it was in August, they had the official signing of the contracts. Everybody met at the school and you were there with your relative, you know, whoever, and then somebody from the company that was indenturing you was there, and you signed your paper.

I remember seeing the people who had all this family support and seeing the ones who didn't have any. As far as the non-minority, just about everybody had somebody—a brother, an uncle, a cousin, a father, a grandfather. There were people with the last names who—you *know* they had three or four plumbers in their family. But with the minorities, I think there was only a couple that had any kind of connections. You could see the looks in people's faces when they would look at each other back and forth across the room. The guys who had all the support are like, Yeah, we're in here, we're in. And the guys who didn't have anybody, you could see that they really felt like they were kind of outcasts. They didn't quite fit into the program because they didn't have somebody in their family who was vouching for them.

And it was funny for me because my family, it really put them in an awkward spot. If I had been a guy and they were vouching for me, I know they would have been totally different toward me. They were being friendly, but it wasn't like "the guys," it wasn't that same kind of comradeship. They were there and they were my family, but it's like, arm's length, you know? It was strange. But once we got all that formality over with, you were on the job and you were in school.

CHAPTER THREE

Crossing the Threshold: First Days

I was as naive as green could be.
—*Lorraine Bertosa, Cleveland, Ohio*

For my very first day in union construction I was sent to a bank in downtown Boston where a journeyman needed a hand pulling wire. Arriving early with my new tools and pouch, I knocked on the glass door in the high-rise lobby and explained to the guard that I was a new apprentice working for the electrical contractor. He refused to let me in. So I sat down on the tile floor, my backpack and tool pouch beside me, and waited for the man whose name I had written down alongside the address and directions on a piece of paper: Dan.

The guard explained to Dan later that he'd figured I was a terrorist planning to bomb the bank. In 1978, that seemed more likely than that I might actually be an apprentice electrician.

My eagerness and ignorance that morning almost combined in disaster. After explaining that I was to push the metal snake, with different colored wires taped to it, into a pipe, Dan left for the location where the wires would be coming out. As vigorously as I could, I began to push in yard after yard of the thin looping strip of metal. About fifteen seconds later, I heard Dan scream. I froze. In moments he reappeared, his ruddy complexion ashen. It hadn't occurred to him that I had to be told to wait until he was at the other end of the pipe and gave a signal, before I pushed a metal snake into a live electric panel—and that pulling wire requires slow, methodical teamwork for safety.

The following day I went out to a "real" fenced-in construction site, bustling with men who worked at trades I'd never even heard of. It was new construction at a chemical plant—an unusually dangerous job, though I had no idea of that at the time. Installations were all stainless steel, as required in hazardous locations. A man from the plant kept *reminding us to run immediately if we ever saw a certain pipe drip. Something corrosive in the environment was dissolving the glue on the soles of my new boots and turning dimes in my pocket green. The crew was no more welcoming.*

—Susan

Women who walked onto construction sites in the late 1970s and early 1980s to begin their apprenticeships in the trades carried more than the usual new job anxiety and disorientation. Most were coming from traditionally female occupations. Not only were the craft and tools unfamiliar, so too were the culture and organization of the construction workplace.

Women used to jobs where two weeks' notice is the norm for hiring or layoff were shocked to learn that, by construction industry standards, a day's notice is considered generous. Those who couldn't afford to risk buying all the necessary tools and work clothes until a job was definite had very little time to get prepared.

A full year after her 1978 acceptance into the carpenters union in Kansas City, Kathy Walsh's phone awakened her unexpectedly at 4:30 one morning. A single parent of three not quite able to manage on her secretarial wages, she was told she had an hour and a half to get to her first construction job.

I WAS LIKE, AH, AH, I can't do that. I mean it wouldn't work with my babysitter and I didn't know where I was going and I didn't have my tools and I didn't want to go off and just not show up at my job. It was like—panic! And he goes, "Well, if you can't do it today, I guess you don't want to be a carpenter."

After a lot of pleading, she talked the apprentice coordinator into giving her a day's grace.

I WENT OUT THAT DAY AND spent every penny I had on a tool pouch. I didn't have boots. Had to buy a pair of work boots and the tools. The list that the apprentice coordinator had given me was very extensive, probably eighty carpenter tools. When you're first starting out you don't know that *these* are drywall tools and you don't need those on a concrete job. And *these* are finish tools. I got as many as I could. I'd been trying to get them all along at garage sales and swap-and-shops. I had my little toolbox and I got my boots and bought a pouch and didn't sleep at all that night.

I don't remember who was babysitting for me at that time, but I either had a friend or made arrangements to take my son early. [I left] at 4:30 in the morning because I didn't know where I was going and I didn't want to be late.

I was so nervous, I was sick to my stomach almost. I had directions to this job-site and I had somebody's name. I didn't know what they would want me to do or anything. I had been practicing what I thought was hammering at home, *tap, tap, tap,* just nailing nails into a two-by-four.

It was really a big thing for me to walk away from the job that I had. The people I was working for were very, very supportive and said, Kathy, go for it, it's an opportunity. If it doesn't work out, you let us know.

Thank God, because I went back to work for them many times.

Women across the country in different trades report similar long waits of six months to a year between acceptance into an apprenticeship and an actual job offer, often while men brought in at the same time went out to work right away. Apprentices usually don't become full union members until they have worked their probationary hours, generally ranging from three months to a year depending on the local. Keeping women in a pool, available for work but not accumulating work hours allowed unions and contractors to be ready to meet compliance requirements without bringing in female members they wouldn't "need" if court challenges to affirmative action succeeded in having the guidelines eliminated. Another factor in this delay likely was that federal regulations bringing women into apprenticeships became effective in June 1978, but the effective date of the regulations for affirmative action on jobs was not until May 1979.

Offering a woman a job—whether she took it or not—counted as a "good faith effort" toward affirmative action. No one checked whether the offer was made under circumstances that made it unrealistic for her to accept, or after such a long wait that she was likely to have found other work. For all the women who did switch their jobs on a day's notice, obviously there were many who could not, and who were recorded on forms as turning down an apprenticeship or a job offer.

Whatever her desire to blend casually into the workforce on her first day, a tradeswoman walking onto a construction site in the late 1970s had all the invisibility of a flashing neon sign. Raised in the Bronx, Melinda Hernandez had become a jeweler's apprentice after leaving college, but found the work exploitative. She participated in a government-funded, two-month training program in construction skills for women that she'd heard about from a friend, and became one of the first four women in the union electrical apprenticeship in New York City. She was not only female, but Puerto Rican and 5'2".

I HAD A LITTLE RED TOOLBOX. I looked like Little Red Riding Hood coming onto the job. It was funny because I had been taught by the women at All Craft, they're going to try to test you, so don't take any lip. You just let them know that you're there to do a day's work and to learn a trade.

Apprentice electrician Melinda Hernandez on the job in New York City.

I was running late because I couldn't find the job. When I walked into the building, I asked for the electricians' shanty. The guys looked at me like, What the hell is she doing here? Like, they heard we were coming, but it was never going to be real, it was never going to materialize. And there is this little woman with a toolbox.

The job was 70 percent complete. The walls were up, the windows were in, and they were just roughing out internally. When I got to the electricians' shanty and I opened the door, the foreman was sitting at the desk. He was about 60 and he was gray, all this silver gray. He looked at me and he says, "Yes, little girl, what is it? Did your father leave his tools home?" He was serious. He wasn't being sarcastic. I think the shop just sent me and didn't inform him, to kind of play this game, Let's see how he reacts, you know.

I was 22, but I looked like 15 or 16 at the time. I said, "I'm here to report for work. I'm an apprentice."

Since construction sites are off limits to the public, most new tradeswomen's first day on the job was also their first glimpse of its sights, smells, and sounds. Just navigating the worksite could be a series of new experiences and challenges.

Barbara Trees had gained some experience with carpentry tools and a taste of working on an all-male crew before beginning her apprenticeship in 1980. As a low-wage artist's assistant, she worked on a replica of the Brooklyn Bridge. Desire for a job that was physical and that paid well enough for her to be a self-supporting single woman without a college degree—and "the momentum of women around me"—drew her into the union.

I DON'T KNOW HOW I GOT this idea, but I actually thought that I was going to be building some nice furniture. I know Mary Garvin at Women in Apprenticeship told me, You have to like working outdoors, you're working in all kinds of weather, it's heavy work, and sometimes it's dangerous. I know they told me all that, but when I headed out for my first jobsite, I was in shock. It was a rude awakening.

I came onto this big jobsite on Roosevelt Island. The tram, this little aerial thing, took me out there. They were building a subway, okay, but I had not a clue. The foreman came up and he shook my hand. It was not like he killed me instantly, which is what I thought was going to happen. There was an empty building across the way. I looked over and thought, It's the only thing that's standing around here. That must be what we're going to work inside of. So I said, "Is that where I'll be working?" He just laughed.

Then we went down into the hole, this dark pit. It was underground. They didn't have it lit very well and it looked just real scary. It was all wet. There was water everywhere. I was afraid I was going to slip and slide down this embankment. I always pictured myself as a pretty strong person and independent. I got into this situation and I was just really afraid. I was afraid to go down a ladder, so I asked him the safest way. I was checking to see if he laughed at me. But he just told me how to do it.

We got downstairs. I remember it was all rebar. They had put the rebar down already. I certainly didn't know that was the name of it, it was all those big metal things. I thought, My God, I'm going to have to walk across this. I was stumbling and thinking I'd go right down through it.

Male apprentices sponsored into their local union by their father or uncle had someone with years of experience making sure they went to their first job properly prepared—mirroring a familiar image. Most new tradeswomen lacked that advantage.

Helen Vozenilek was a strong athlete and tired of the minimum-wage jobs such as warehousing she'd held since dropping out of college. Finding a notice on the bulletin board at the YWCA about openings in the electricians apprenticeship, shortly after arriving in Albuquerque from the Northwest, was "pure luck." On her first job she improvised her tool pouch as best she could, and tried to stay calm when the foreman's brief jobsite tour included taking her up to the roof.

IDIDN'T LIKE HEIGHTS. I MEAN, we're not talking high at all—it was two stories—but I get to the edge and already I get that feeling like I was ready to throw up. At high places I get that sudden urge to jump, too. I don't know if other people get it, but it's split-second and then right away—it's not very serious at all. All those things would be going through me and I thought, Jesus, what am I doing here? I'm only two stories up and already I'm nervous.

On that tool belt, I didn't know where the tools went. I didn't know anyone that was in the trade to ask except this one woman and I wasn't going to call her up and ask her which pocket the screwdriver goes in! You know, they have the little thing that you hook the electrical tape on? Ah, geez, I didn't know— I was hanging tools from that little thing! This guy out there helped me set up the tool pouch and it looked better, but it looked obvious. You can tell a new tool pouch. Shortly afterwards, I backed a car over it to make it look used, because you just get too much ridicule.

The boss paired me up with this other guy who was real sweet. I didn't know anything, didn't even know which direction the toggle bolts go in. He showed me how to do something and I kept trying to suck up this toggle bolt. But I had the toggle going the wrong way! Of course, the thing wouldn't suck up and the hole was getting bigger and bigger in the sheetrock and Oh, my God!

Females in our culture learn early to avoid isolated situations with men. Yet most tradeswomen were put on jobs where they were not only the first but the only woman. Cynthia Long had met Melinda Hernandez as she waited on line—five days and nights camped out on the streets of New York—for an application to the electricians union apprenticeship, and again when new apprentices had been called into the union hall for an orientation and tool check. But Cynthia walked onto her first job alone, with a fear and strategy familiar to most women.

IHAD MADE A CONSCIOUS DECISION that I wanted to be perceived of as a professional person. I had my little work uniform from Sears and Roebuck, the dark blue chinos pants and the work shirt. I went dressed to work. My hair was pulled back and up out of the way so that I would have free movement of my head and not worry about getting my hair caught on something in the ceiling. I remember my pliers were Ideal pliers—they were amateur tools, the only stuff that I had.

I was scared. First there was this whole thing about finding the shanty. I had never, ever been on a construction site per se and certainly not the size of the construction sites that are typical here in New York. I didn't know where to go. One of the concerns high on my mind was my safety, was I going to be safe—in terms of rape. That clearly crossed my mind, Is this going to be something that I have to be concerned about and constantly fearful of?

Rumors skipped around the country via women's movement networks. Sara Driscoll, who also started her electrical apprenticeship in 1978, walked onto her first job with trepidation.

MY FIRST JOB WAS DOWN at the South Postal Annex, downtown Boston. Humongous site. I mean, this building went on for about three blocks, it was huge. And the first day I walked on that job I was really frightened. I wasn't frightened of the work. It was, What are these guys going to do to me, are they going to even let me be here?

I had heard some stories about some women in the trades down in the South. I remember one that really stuck in my mind was this woman in Texas who was a carpenter. They pounded her hands with a hammer, they broke her hands, that kind of stuff. I don't know if it was a true story. I don't remember where that story came from. I walked onto that job and there were probably 150 tradesmen from all the different trades, carpenters and tin-knockers and pipefitters and electricians and what-have-you's. And I was the only woman. And it was this humongous job.

Apprentices are supposed to be paired up with an experienced journeyperson whom they are required to follow into isolated mechanical rooms, poorly-lit basements, or anywhere else there is work to be done. Many situations normal to a construction job ring a caution bell for women. Deb Williams was 17 when she began her painting apprenticeship on a job at Harvard University.

THERE WERE SO MANY MEN on that job, I was afraid. I didn't want to get caught anywhere alone with any of these people because I didn't know them.

One guy, coming back from lunch, he's talking to me, Hi, Deb, how are you? How was your lunch?

Fine.

When we got into the lobby to take the elevator up to the department that we were working in, the elevator doors open. He got in and I stayed there. And the elevator doors closed—because I didn't know this person. I was damned if I was going to get in an elevator with a strange guy, just him and me—no way! To this day, he's my best friend and he'll tell you the same story. He goes, I got in that elevator, I'm saying, do I have B.O.? What's the matter? Did I do something to her?

I was leery of all men.

A short while later Deb was transferred to a bridge being sandblasted and painted.

THERE'S ONLY ABOUT TWENTY MEN on that job, which is totally different because now they're all grubs, they're all sandblasters. They're not all these nice white guy painters that you see on the label of paint cans. Now they've all got hoods on, it's freezing out—these are some rugged people.

We was hanging tops and when I was climbing on the steel girders I was on one side and my foreman was on the other side and the wind is blowing and I'm doing all I can to hold this thing. I'm not as strong as he is and the thing is ripping. And so I hooked my side. He said to me, "Deb, climb over me and get on the other side and hook it up." He's holding it in the middle. We're on our bellies on a girder and there's another girder about two feet above me. And he's telling me to slide over him across his whole body and get on the other side and hook it up. I said, "No."

He said, "Deb, you got to slide over me. It's the only way to go over there."

I said, "No." 'Cause I had to wedge myself between the beam and him. My body had to slam against his to get over. He says, "Do it fucking now. I'm holding this thing. Slide over me!"

And whoosh, over I went and hooked it up. He goes, "Now was that so damn bad? I could have fallen off there. You're here to do a job."

My face was bright red. Even thinking about it now.

The situation was brand-new not only for the tradeswomen, but for their male co-workers as well. Not only had the men not worked with a woman on a construction job, many had little experience working with women on any job. For those with backgrounds in vocational schools and the military, the single-sex culture they'd spent their days in since puberty had suddenly been changed by federal regulation.

Men's discomfort with women's sudden appearance in their workplace took on many forms. To survive, tradeswomen quickly had to become stand-up improv artists with a flexible repertoire of responses. Maura Russell, a graduate of Smith College, began her plumbing apprenticeship on a medical research building in Cambridge, where men's reactions ranged from hostile to protective.

ON THAT FIRST JOB THERE were a couple of guys who, whenever I was within earshot, would get into this loud and graphic and disgusting rap about all the horrible things they had done to women. Glance over to see that you're there and kind of check out the reaction. I don't even think I knew their names. They were electricians. That was just the environment.

Definitely the kind of job [where] when you went out to the coffee truck— complete silence. You were really, really, really the oddball. Big time. Walking out of the building and going to the coffee truck and getting total silence—I mean conversations stopped completely and people stare at you the entire time you're there—that was bad.

I remember the first day on that job. It was dark because I wanted to be there early, make sure I wasn't late. It was fall. This little old stooped laborer met me and just said, "Oh, no. You shouldn't be standing out here. They're all going to be coming and they're all waiting for you." It was so scary. He says, "Come and wait over here."

He takes me to this little shed with no lights and closes the door. He pokes his head back in just before he leaves and says, "You know, they're waiting for you, but they're all going to be surprised. They're all wrong. They thought it was going to be a black girl." And then he closes the door.

I don't know what time it is, whether I should go out. Oh, scary as anything! I stayed in there for quite a while and then when I finally got the courage to open the door to God-knows-what, it was light. I was actually a little bit late at this point. Probably people had been working for five minutes or something, and I have to go out and find out where my boss is.

Like characters in a slapstick film, tradeswomen often wore the wrong amount of armor for the situation. As soon as they added or removed some, though, the circumstances changed, and they were wrong again.

The morning after being awakened by the 4:30 am call, Kathy Walsh began her apprenticeship. She put her new carpentry tools in the car and followed the apprentice coordinator's directions to an artificial lake under construction.

I DROVE UP TO THIS PLACE and I had no idea that it was going to be this huge—this is a bottom of a lake!—and all the excavators and all the heavy equipment is out there. I saw some people, so I drove over there and I asked for this guy. These guys said, "He was just here. He just left and he went off that way. Follow those car tracks in the road over there and you'll run into him." So I did that and I ran into another group of guys that were getting ready to go to work. They said, "Oh, no, we haven't seen him yet this morning. Usually he's up over this hill in that trailer about now."

There's this huge open area and I'm driving. I had a little hatchback Mazda, I'll never forget it. It was bouncing up and down over these ruts. I went to about eight or nine different places and they're all saying, "No, don't know him." "Oh, he was just here, go over there." I found out later that they all knew that there was a green woman coming on the job that day and they were all fucking with me.

I finally find him. I'm so flustered at that point. It's like an hour after the job was supposed to start. I'd been driving around for two and a half hours trying not to be upset and I don't want to cry.

He's standing there. I'd already talked to him through the car window so I knew I was at the right place. I get out of my car, take hold of the top of the door, and shut it with my hand in it. I'm like, *Oh my God.* Open the door with my other hand, take my hand out. Down to the bone on all four fingers! It's bleeding and I'm trying to hold my hand behind my back and keep him from noticing the blood running down onto the ground. He says, "Park your car over here. See those people down there? Get your tools. Go see so-and-so and do whatever he tells you to do."

Thank God I had some tape in my toolbox. I taped up my fingers—I mean, they were just hamburger, all four of them, down to the bone—and went to work in a mud hole. We had to go down into about a forty-foot embankment where this drainage system was going in. There were only about ten guys working in this area. One of them introduced himself and was real friendly from the very start. One of them, a huge ironworker—I'm 6 foot tall, this guy had to be 6′8″—said some comment about, he couldn't believe there were *fucking women* out on construction jobs now. Made it very clear that he did not want me out there.

Out in Berkeley, California, Paulette Jourdan still carried a childhood dream to work with her hands and build the squirrel cage from sticks and string she'd imagined. In the mid-seventies she had looked into apprenticeship opportunities with the carpenters union, but didn't even apply once she realized that the starting pay scale was too low for a divorced mother of three. In 1980, when her children were older, she signed up for the Plumbers exam when a friend heard about openings on the radio. For Paulette, the choice to become an apprentice plumber was a major decision toward personal empowerment. On her first day on the job, though, not much seemed to go right.

THE DISPATCHER CALLED UP AND said, "We have a job for you and you can start school. You need to go out to Livermore Lab." They were doing tricks with the affirmative action thing. They needed not only a green apprentice but it had to be a woman. The extra icing on the cake—I was a woman of color. That was an advantage for them. They got a very cheap worker and they fulfilled the two requirements of affirmative action, so they only had to hire one instead of two.

Well, I didn't have a car. Livermore is like forty miles from Berkeley. I called my sister and asked if I could borrow her car.

I go all the way out there, I get through the first gate, and the guy stops me and says, "You have to go into the office and check in." Livermore is a secured lab, you have to have security clearance to even drive through the gate. They ask for my driver's license—and I don't have it.

I left my driver's license in my cigarette case at a friend's house. She moved to Sacramento without telling me, called me up a week later, and said, I have your cigarette case here. I said, Oh, mail it to me. She didn't get around to it. My car was broken. I had been taking the bus, doing odd jobs like cleaning house so we could survive. I didn't have a car, so I didn't worry about the driver's license.

The next thing I know I get this call. I didn't even think about the driver's license.

They won't let me check in, won't even let me go to the assigned building inside the complex. They made me turn around. I had to drive all the way to Sacramento, seventy-five miles one way. I was falling asleep on the freeway, it was crazy. I got the driver's license, drove all the way back to Livermore. I was afraid if I didn't show up they were going to kick me out of the program.

It was cold, it was wet, it had been raining. It was the end of November, the rainy season. They gave me a hardhat when I checked in at the trailer. I had to walk through the mud to get over to the main building they were working on, about 200 yards away. One apprentice and four journeymen. I was going to be the second apprentice.

It wasn't multi-storied, it was a cathedral almost, 30-foot ceilings, because they had to run so much pipe everywhere. They were running 4-inch cast iron and 4-inch and 3-inch copper pipe for these huge systems, air and water and gases. They're on these huge scissor lifts.

I think the first thing that blew me away was how filthy everything was. There were electrical cords everywhere—you'd trip over them. There were pieces of metal just strewn, I'd never seen anything like it. I was blown away by the immensity of it all and it just seemed incredibly disorganized. I don't know how anybody got anything done.

I get there and I don't know a pair of Channellocks from a piece of pipe. I don't know anything. I didn't think there was anything wrong with that. Nobody said I had to know anything. Nobody said you need to be prepared to do this work. I passed the test. They said, Go to work. I get there and these guys, they sort of look.

I'm scrawny. I had this jacket and these overalls that were too big and these clunky boots. These guys did not want to see me. They were working, so only one or two came out at first. I was standing around waiting for them to tell me what to do. Finally I asked one of them, "What do you want me to do?" I don't remember what he said, but it was very vague. I said, "Aren't you guys supposed to train me?" I wasn't nasty, I just asked a question. I was just trying to find out where to start, what I was supposed to do.

They looked at me, I think there were three of them, and two of them said at the same time, "We ain't got to show you shit." And they turned around and walked away, left me standing there, just hanging around the gang box area, not really sure if I should go back to the trailer or hang around and wait for somebody to come through. *We ain't got to show you shit.* I didn't see them again for an hour.

CHAPTER FOUR

Ain't Got to Show You Shit

I have had a lifelong interest in mechanical things, but not the appropriate experience.

I had this faith that the apprentice system would take me from nothing and teach me everything.

—*Marge Wood, Madison, Wisconsin*

The job was a new forty-story office building under construction—no windows against the Boston winter. The crew foreman was a gentlemanly old-timer with dancing blue eyes who in his prime had run many large construction projects, a real sweetheart. Despite the fact that another foreman had introduced me to him my first day there as "something to keep you warm," Bob always treated me with the same respect and playfulness he showed everyone. I was a third-year apprentice, my wage scale had just taken a big jump. Although I did my share of apprentice duties, like getting the crew coffee and bringing up stock, Bob made sure I also had time to gain experience at the trade. Which really irked the general foreman.

When the other apprentice was pulled off our crew, my coffee duties doubled and Bob was told to have me do only housecleaning tasks: keep the stock neatly arranged, sweep up, and make sure unused supplies were put back away. But whenever he could, Bob would get me working with the tools. The general foreman one day on his walk around the job caught me laying out a bank of 2-inch pipes with a journeyman. Within a few minutes Bob came by to tell me I'd have to put away my tools and check around for loose stock.

Next morning, a first-year white male apprentice appeared on our crew. He was assigned to that pipe run—and I was told to add his coffee order to my list.

By tradition, when more than one apprentice is on a crew, either the drudge work is evenly divided, with duties like the coffee run switched off daily or weekly, or a strict pecking order is followed, so that, for example, only the first-year apprentice does the go-fering (go-fer coffee, go-fer stock). When it became clear that no switching off was intended, I felt I had to complain. According to Bob, he'd been ordered by the GF to keep the first-year on the pipe job, and me, a third-year, on housekeeping. The coffee run would not be rotating because the other apprentice was "not really on our crew," he was "on loan" from another crew. I'd never heard of such an arrangement. Nor had Bob. So the next school night I spoke to the apprenticeship director, who told me to bring the problem to the job steward.

I tracked him down after morning coffee. The steward shook his head and said that if I'd gotten the morning coffee, the other apprentice would get it in the afternoon. I considered the matter settled.

That afternoon I didn't go for coffee. Neither did the other apprentice.

The general foreman's beet-red face was screaming into mine first thing that next morning. Not only was I interfering with the construction of a forty-story high-rise, he told me, but he'd seen Bob carrying stock up the stairs rather than assigning it to me. When Bob had a heart attack, I'd be responsible.

I made it clear that it wasn't my intention to kill Bob, disrupt construction, or deny anyone coffee. But I was there to learn the trade. And I was following the steward's advice in standing up for my rights.

Next morning I was splicing wires when the general foreman walked by. "How are you doing?" he asked. "Okay," I replied warily. "Just okay?" he blustered. "You're working with the tools, you should be doing great!"

"I'm doing great," I said.

—*Susan*

Apprenticeship training combines two key components: classroom instruction and on-the-job training. Union apprenticeship programs are generally run by a director or coordinator, often a member with teaching certification, and governed by a joint industry board representing the local union and signatory contractors. Apprentices work under the union's contract and are indentured, signing an agreement with a contractor exchanging labor for training. The length of the apprenticeship varies, from three to five years, depending on the trade. Step raises increase their pay automatically, usually as hours worked accumulate, so that a first-year apprentice may start at 30 percent or 40 percent of the journeyman's rate and reach 80 percent by fourth year. For contractors, low-rate apprentices are essentially cheap labor who do the menial work that allows highly paid journeymen and -women to be more productive. Ideally, during their training time apprentices take on increasing responsibilities and develop the wide range of skills, judgment, and confidence necessary for them to graduate as competent journeylevel mechanics.

With trust in that process, Paulette Jourdan stood alone outside at the Livermore Labs on her first day as an apprentice plumber until

Finally, somebody did come through. Somebody assigned me to him. He hated it. I was too skinny, I wasn't strong enough. He had me up on some platform cutting copper pipe. And half-inch copper is very hard to hold on to, you have to crook your last finger to keep it steady while you cut it with the cutters. He showed me how to do it, but I wasn't used to using my strength—if I had any. And I was taking a long time. They had to show me what the tools were—they were pissed about that.

The first day I didn't feel too bad. I didn't even mind being called Stupid, 'cause I knew I didn't know anything. It was the second day, that week, when it started to build up. This man—he was a skinny little guy himself—he was so pissed I was there, he started calling me Josephine. I got it instantly, as soon as he said it—that's what makes him comfortable. I kept thinking in the back of my head, If I don't make this work I'm going to consider myself a failure. I let him call me Josephine. He got a kick out of that. I didn't care—just teach me. And that's sort of how I got through the entire apprenticeship. I think I probably compromised myself more often than I would have liked to, just so they would train me.

Since there was no first-year class in her local when Paulette started, she was placed with the second-year apprentices for classroom instruction, and moved along with them until her fourth and final year, when she went back and studied the first-year material. At the midway point in her apprenticeship, she assumed her training was on track and that "whatever pieces I don't know, eventually I'll learn."

I just finished my second year, which means they call you a third-year apprentice but you aren't, it's the *start* of your third year. I was out on a jobsite, they were putting in underground pipe, and I saw these pipes sticking up. There was nothing there but dirt and a few little trenches. The carpenters had just set up some batter boards and pieces of string. I was cutting pipe for this journeyman and bringing him fittings and I was very curious, I was starting to notice things. I looked over across the field and I said, "What are those two pipes sticking out of the ground?" And he looked at me.

He was down in the trench. He said, "What year are you?" And I said, "I just finished my second year." He said, "You're a third-year apprentice and you don't know what the hell we're doing?" Then I really got nervous.

I didn't realize that they were gonna pour a concrete slab, that's why the pipe was going into the dirt. It was a residential project, it was gonna be a two-story house. All he said was, "One's a kitchen, one's a laundry, and that one's for going upstairs." There's three pipes sticking up out of the ground about two feet and I'm thinking, Oh God.

He throws this drawing at me. It was an isometric drawing and I didn't know what the hell that was. I just saw a bunch of lines. All that day he spent trying to tell me that I should have learned this stuff, he didn't have time to train me, they were watching him. About lunch time I saw the project coordinator coming toward me and sure enough, white envelope. Handed me the envelope. I thought I would die. You know how your mouth fills up with salt just before you cry? I couldn't even talk. The guy handed me my check, he said, "I'm sorry. You don't know enough. You need some better training." Oh yeah, I need some better training, but there's nobody who wants to give it to me. I didn't say all that but that's what I was thinking. That's the excuse they always have. They don't want to be the ones to spend money to have one of their people take enough time to train you.

I went to the union hall. I thought, That's it, they're gonna just kick you out now, you don't know shit. It was the first time somebody had said to me, You don't know enough, and I had no idea how much I was supposed to know. I checked in at the counter and told them I was laid off and I'm sort of wandering in a daze, What am I going to tell my kids? But for some reason I think, Let me go see if the school coordinator is in his office.

I had no idea I was going to confess to him. "I'm laid off," I said. "How much am I supposed to know anyway?" He sort of looked at me and he said, "What do you mean? What kind of work were you doing?"

"Residential. There were some pipes sticking out of the ground, I didn't know what the hell they were. How much is a second-year or a third-year or fourth-year supposed to know?" He started scratching his head, "Well, it depends on if the journeyman takes you under his wing," and "Some people learn fast," and all this bullshit. I said, "Wait a minute. I just started my third year, how much am I supposed to know?" He said, "I don't know."

I couldn't believe it. I said, "You guys don't have an agenda?" I really got mad. So he calms me down, he says, "Pull up your chair." He drew a picture of a bathroom with the three fixtures in it and he said, "Now draw the sewer line." I drew the sewer line and he said, "Now where are the vents?"

And I didn't know what vents were. I'd heard the term but I didn't know what the function was. He said, "Everything has to be vented through the roof." And so I watched him. The toilet pipes extended halfway into the middle of the room and he drew a line back to the wall. I said, "How come you're drawing lines over there?" He said, "You can't bring the pipe up through the floor in the middle of the room. You have to run it back over through the wall and bring it up."

A light went off. Something clicked.

I think we spent about three or four hours in his office with him drawing pictures and us talking. It was like *The Miracle Worker* when the deaf girl finally understands what all those finger movements were about. And then he turns around and says—like it wasn't anybody's fault—"That's why you should have started on residential."

Like, How come you didn't put me on residential?

Residential let's you see a whole picture within a month. You get to *see* where things originate from and where they go to, *why* the water lines are run this way, the waste and the vents and all that. I was on commercial for two years. I never saw a whole of anything. They were running pipes, I didn't know what the hell they were for.

I did thank him. He could have done that all along, but he didn't know I needed it. 'Cause I didn't know to ask. After that I was fine. I went home and studied my books. I went on the next job, I knew what things were. I started to understand.

Classroom curriculum was generally set at the International level of each union, to standardize instruction for that trade across the country. In reality, though, programs varied widely. In one local, school might be viewed as an essential element of the training, while in another, just a paper requirement. An apprenticeship coordinator could be selected for his skills as an educator; or for his political connections. Some locals could afford modern training facilities while others were inadequately supplied. "When I was taught how to hang a door and install finish hardware," Kathy Walsh remembers, "they didn't have any doors at the training school. We put hinges on a two-by-four and hung a two-by-four and pretended we were hanging a door."

Most tradeswomen, such as Los Angeles ironworker Mary Michels, found more equality in the classroom than on the job. "In the classroom I would say I got excellent teachers. I think they went out of their way to make sure I understood the lessons. The apprenticeship school, I have very pleasant memories." Women who were isolated, one to a jobsite, could meet each other and develop friendships and support systems at school.

When the instructional programs were lax or farcical, however, they not only felt like a waste of time after a hard day's work out in the weather, but they tended to foster an atmosphere that was particularly uncomfortable for women. As an apprentice, painter Yvonne Valles would often ask to leave the classroom and do her studying off by herself in one of the training areas.

I'D HAVE TO GO TO school every three months for a whole week. I'd have to take off work and get on unemployment and go to apprenticeship school because that was their policy. The instructors, a lot of them, they couldn't care less about the quality of training. The guys in the class, there'd be maybe thirty, a bunch of youngsters, a lot of them high school dropouts. They'd be real unruly, they'd be yelling, they'd be making sexually explicit comments. I'd turn around and tell them to knock it off. Sometimes they'd start heckling me and I'd have to walk out and call the instructor. It wasn't always directed at me, it was mainly just that they talked like they were in the locker room. They didn't care if I was there. We'd have to study all these chapters during that week. For like three days we'd be in the classroom. You tell them to be quiet and they're quiet for a few minutes and then they start in again, they're throwing papers at each other.

Then they'd give us a test. Well, the guys had a cheat sheet that they'd pass around, and next to every answer—they were usually multiple choice, or true or false—they'd put a little pinhole. That's how most of those guys would pass. They'd pass me the cheat sheet. I'd say, I'm sorry, I don't want it. I'm going to learn this and I'm going to read this book and I'm going to know what this is all about.

Another thing that used to happen at the apprenticeship school was one of the instructors used to let the guys bring in porno movies. He'd just say, Read this chapter, then we're going to watch a movie. If it was during lunch break he'd let them watch porno movies. It was no secret, everybody knew that that's what was going on.

The only part I really liked about it was that one of the instructors, he was from Minnesota and he came from a line of master painters, some of them were from Europe. His family and he can do all this real fancy stuff like marbleizing, gold leaf, sign painting. He'd let us bring in antiques or collectibles or whatever we had at home. When we had our classes in wood graining or refinishing or marbleizing, he'd let us bring in projects. That was a part of the whole apprenticeship program that I really liked because I like real creative work. I'd look forward to going to school when I knew I was going to take a class with him in wood finishing. But the rest of it was boring.

Donna Levitt served her carpentry apprenticeship during a transition.

THE FIRST THREE YEARS I was an apprentice we went, I think, twice a week at night. There was a bar across the street. We spent half the time at the bar. The teaching was lousy.

The last year, we switched to daytime training one week every three months and there's a state-of-the-art facility. I actually learned something.

Even the best classroom instruction could enhance but never replace the practical skills learned on the job, where subjective factors played a larger role. When a job ends or a crew downsizes, the contractor can either transfer workers to another job or lay them off. Depending on the local's policy, unemployed apprentices either found their own next job or were assigned to one by the apprentice coordinator. Ideally, apprentices would work on a series of jobsites that exposed them to different aspects of the trade, or spend a long time on one job that offered a range of skill training, as Diane Maurer did in Seattle with her fourth electrical contractor.

EﾟND OF MY SECOND YEAR I went to work on a seven-story office building. From the ground up. I was the second person there after the foreman. We ran the pipe for the lights and the elevator pit. I worked on the building until it was finished. I felt like I learned fairly quickly and I was willing to do whatever jobs were put before me and, you know, I didn't complain. I really did feel like I had a good cross section of journeymen, and everybody always was real patient. I don't remember ever working for anybody that ever yelled at me and called me names or anything. The foreman saw to it that I worked

Electrician Diane Maurer of Seattle, the year she turned out (1983).

on every aspect of the job, from the switchgear on the main feeders to the risers in the electrical room to the underfloor duct and working the ceiling, which was really good. I really appreciated it and I couldn't ask for anything more fair than that.

I had heard a lot of women in the apprenticeship program complain about the type of training that they were getting. I'm sure it was a combination of things. I ended up there, whereas other apprentices ended up on high-rises and they were doing the same thing for months on end. Working on a job, any job, from the beginning until completion is a good experience for anybody. I don't think a lot of people get that opportunity. I worked that shop until I turned out. I had good training.

The economic situation in the city—what was actually being built or renovated—defined the job opportunities in that area. Many women were further restricted to those contractors willing to hire females. The low number of women taken into the trades, encouraged contractors to place and keep them on jobs monitored for affirmative action. All of these factors limited a female apprentice's opportunity to gain the full exposure and training she would need as a journeylevel mechanic. Mary Michels found, "There's a lot of companies out there that just don't hire women, period. As an apprentice, you're allowed to hustle your own jobs. The big companies that put up the structural, they've never hired me, so I have gotten with the pre-cast companies."

As a carpentry apprentice in San Francisco, Donna Levitt's experiences were similar:

IT WAS VERY DIFFICULT FOR women to get, for instance, residential framing. There are very few of those jobs that have government funding, for one thing. And they're mostly out in the suburban counties. It's more of a cowboy mentality, the residential framing jobs, and women are not included. In my apprenticeship, I really made it a point to try to get a well-rounded training. The program does little to encourage that. It's kind of left up to you. If you go through your whole apprenticeship doing form work, so be it.

Carpenter Kathy Walsh found that while many of her male counterparts had a relative who could use personal connections to get them on jobs where they could learn a range of skills the same was not true for women.

FOR THE GROUP OF WOMEN that first went through, we all got pigeonholed into certain areas and that was where the coordinator kept us. To give him the benefit of the doubt, I would say that he was trying to give us an area of specialty where we could get a lot of experience.

I always did form work. Most of the girls did form work, because the Department of Transportation of course was the only area that they really even looked at the number of women and minorities out on jobs. So doing bridge work and form work—a lot of the women got pigeonholed into that. Later on in my apprenticeship I was like freaking out, going, Oh shit, I'm going to turn out and be a journeylevel carpenter and have to compete with these guys for these jobs and this is the only thing I know how to do.

I tried several times to get different types of contractors to hire me, and I never did get anybody to do that, but two different contractors actually called me back to work more than once when a job was finished. With both of those contractors I was able to get just a little bit of real on-the-job experience hanging cabinets and doing finish work for short periods of time.

But boy, I could build you anything out of concrete!

Ironically, the jobs to which women were ghettoized tended not to be the ones requiring stereotypically female fine motor skills—finish carpentry work or ornamental ironwork—but the duller, heavier tasks. Although she herself got a well-rounded training, many of the women who began their ironworker apprenticeship with Randy Loomans did not.

I STARTED OUT WITH A CLASS of five and I was the only one that finished. I'm thinking, Why? Was it so hard? Did they beat them down so bad in the rod patch? If the coordinators would do their jobs, that wouldn't happen—to make sure that no woman has to spend more than six months in the rod patch. That usually is where the need for women most was, because it's highway money or state money or dams, or somewhere where rods were a big part of the contract and they had to have women. Then a company would get a woman, and they wouldn't turn her loose. It's almost in our apprenticeship contract that we get so many hours of rods. And once you have them hours, it's up to the apprenticeship coordinator to pull you out of that and send you on to a structural, to a sheeting, to a miscellaneous. Finish work's a fabulous part of the trade. If you get your six months in rods, why do women end up being there for three years? or two years? or even a year? That's too much. That would discourage anybody.

I think about rods and I think that's the most back-breaking work I'd ever done. Hauling it out on the decks, big eleven iron, that's *oooh*. And it was the most defeating, in the sense that here you've worked your ass off to tighten up this mat a hundred percent—then all of a sudden they come in with the concrete trucks, pour it over. That's why it was defeating to me. Because your blood, sweat, and tears is under there and nobody knows. I have to see results, I guess, of my work.

Mercedes Tompkins, who left a job repairing office equipment to become an apprentice electrician in Boston, concluded from her first job that for her the training was too dependent on chance. Apprentices, unlike journeymen, did not have the right to quit a contractor. Quitting a job meant quitting the program.

IT WAS A CREAM PUFF JOB, installing hard-wired smokes in the Prudential Center. We were in people's apartments. It was early September. You could stay the whole winter and not have to deal with the cold weather.

I lasted probably about six weeks.

As soon as I came on the job, the pin-ups went up, the naked ladies and the jokes—the works. The first day I was told to move all this conduit from one garage to the next. The next day I was told to move it back to where I had taken it out. Back and forth, back and forth.

I was just this woman and they didn't really know how to deal with me. I would not overrule the fact that they may have had some racial overtones, but I think that these people would have treated any woman poorly, I really do.

There was a young white guy, an apprentice, who came on a couple days after me. It was his second day, he didn't know shit. Except his father was an electrician. He was learning from Jump Street. They gave him work to do. He wasn't running for the coffee, I was running for the coffee. I was just to clean up after—vacuum and stuff. He didn't have to clean up.

I was told by my foreman that I could not leave any tools or any equipment in the hallways or anywhere. If they dropped the tools I was supposed to put them back on the cart. If they left their belt out at lunch, I was responsible for putting their belt on the cart.

One day they all just sort of left. They didn't tell me that it was lunchtime, I didn't know why they were leaving. They left me with all of the tools. I felt like I couldn't leave. A half an hour later they were back. I said, "I'm going to lunch." They all started laughing. "Well, you can't have lunch. You missed lunch, honey." I said, "But I'm supposed to be watching your tools." "Well you know, that's how things go." I went down and talked to the foreman, "Listen, I didn't have lunch today. What's the story? The guys left me with the tools." "Well, you're responsible for the tools. That's right."

Then they put me in the mechanical room, like on the fifty-third floor in a totally different building across the complex, installing some piece in the detectors. Benchwork. Sitting up there by myself screwing in these parts with the air-conditioning *rrrrr, rrrrr, rrrrr, rrrrr.* I'm feeling like an alien. I mean, What am I learning? I didn't see signs of them being invested in me as a person and my skills development.

I felt really unsupported by the women, the white women, who I perceived to be my support network, because how they took on the oppression was totally different than how I perceived it. "Well, you know that happens on the job. It's just because you're an apprentice." Sort of, you have to take it. But as a black woman, what did it mean for me to be subservient to these guys, to be an indentured slave—even the language!

It was really affecting my self-esteem. I was questioning my own judgment: Is this supposed to be happening? Should I just put up with it? Shouldn't I put up with it? Maybe I'm just being a whiner.

And the pay was shitty. The system was based on you being 18 years old, living at home, and all you needed was spending change, $95 a week. They expected you to live off that money! And to produce in school and to produce on the job! A lot of the women had families, the black men had families and responsibilities to keeping those families fed and clothed and housed. The system was just not logical.

I realized that in fact Xerox [where I worked before] wasn't that bad. Xerox's premise is to get you up and running in the shortest amount of time. Two months of training and they want you out there in the field and doing your work. They would compensate you in terms of pay if you were more productive.

The union's premise is, they want to slow down the process. Cheap labor, four years of this whole gradation—your first year you do *this* and it gets a little better. Well, it's too random. It depended on what jobs you were on. It depends on the foreman and how willing he is to train you. And if he has a program, he's organized. If I was going to be on *this* job for a year, not learning anything, it didn't really make a whole lot of sense.

I said to my foreman, "I don't want to do this anymore."

The company called me in to talk. I think they were a little bit concerned that I might sue. I had no contact with any of the union officials. I just sort of bowed out.

If I had gone into the union before I went to Xerox, I probably would have stayed there much longer. But I had already been of value to some other company, some other industry. Why would I put up with that shit? I came in feeling, Oh, union, union, I want to join the union! I left it feeling real disillusioned.

The heart of an apprentice's training—just as it's been for centuries—is the experience of working one-on-one under the guidance of a skilled tradesperson. In the Middle Ages an apprentice often lived in the home of the master to whom he was indentured. Apprentices in modern craft unions were traditionally the sons or nephews of journeymen in the local, often working for the same shop that a relative did. Female apprentices, brought into the industry through the actions of the federal government, were still dependent for the core of their training on their relationship with their supervising journeyman on the job. The quality of their training was in many ways determined by the quality of that relationship. As an apprentice in Albuquerque, Helen Vozenilek found that much was left to the journeyman's discretion.

As AN APPRENTICE THE WHOLE structure is set up like a master/slave relationship and it sort of depends on how benevolent the master is and how they treat you. Some journeymen don't like the relationship that way and tried to make it more a partnership. Some journeymen do like that strict hierarchy. They bark at you to do something and you do it. Whether the slave is male or female—that does have a lot of bearing—but the whole inequity of how the relationship is set up makes it possible for one not to learn something, to be told, You have to do this, and not being given any background.

The withholding of knowledge is very powerful. I would be around guys a lot that wouldn't explain why they did things. After a while, it gets real hard always saying, Why did you do that? Why? Why? Why? Why? You feel like an idiot and it's self-perpetuating. You're not encouraged to ask, you're not encouraged to learn, so you don't. And you stay at the same level of understanding. I remember pulling wire with this one guy in this whole senior citizens [complex]. At every outlet he would say either, Leave the neutral long, or Pull them up tight, or Leave one of the hots long. He would tell me which colors to use. And he'd never explain why. Obviously, you leave the neutral long because you're going to pigtail it or something. None of the why's behind things were explained to me. Hardly ever. Those were the good journeymen that would tell you why you did stuff.

Larger shops often had different training tracks for different apprentices, depending on whether they were considered "foreman potential" or "short-term cheap labor." A father or uncle familiar with the shop could help open opportunities or advise his son or nephew on how best to handle a given work situation. Although an apprentice, by union contract, was required to work under a journeyman's supervision, some women had to speak up even to be given a partner or work that involved training. Bernadette Gross went through her carpentry apprenticeship in Seattle.

I GET THERE AND THE GUY has me sweeping out the buildings after the painters have been there and everybody's been there. I got to him and I say, "I'm an apprentice carpenter and I would like to work with a carpenter so I can learn." I'm a mild-mannered person, I've never been really aggressive, but I step up to him and I say I had this right. He was huge, with red hair and a red beard. I'm 5'3". He's about 6'2". He said, "Yeah, well, keep on sweeping." I'm talking to these other women and they're saying, Well, go to the shop steward. So I go to the shop steward and I say, "I'm not being used in my proper capacity. I'm being used as a laborer and I'm an apprentice carpenter. He has me sweeping all the time." And he says, "Tsk, tsk, tsk, he shouldn't treat you like that." I couldn't think of anything else to say, because I am very good at knowing when I'm dealing with a brick wall. I'm not one of those people that just like, ram into it anyway. But I will follow the channels. So I go to the superintendent and I tell the superintendent that, you know, I'm just not being used as an apprentice.

I come to work the next day and the foreman says, "So, you want to be a carpenter, huh? I have a job for you." And he gives me—it's the dead of summer—cutting all the plaster seams. Which means I have to have on my goggles, this World War II gas mask-type thing, gloves, I mean, completely covered from head to toe, and I have to cut all the seams with the Saws-All. A lot of the seams had metal mesh.

So I get in there and I'm going to cut the walls the best I can. In two days, I was cutting up a storm and the fellas came to me and said, "You're going to have to slow down. You're going way too fast. You're going to have to learn how to pace yourself." And they began to show me some things. Next thing you know, I was on the bathroom framing crew, and when they were doing the porch, roofing, and the siding, I sort of just moved along.

Even with speaking up, it was [the same] with every job. I picked my battles. If I was on a job that there was really a lot of carpentry work being done, then I wanted to be a part of it. If I was on a job where there was really this borderline labor stuff where you're doing retaining walls and that kind of stuff, I just hung in there with that.

Without mentors, tradeswomen and other non-relatives often had to be more assertive and develop strategies through trial and error to ensure their training. A journeyman's foremost concern, after all, was his own job security. For eight months of her first year, New York City apprentice Cynthia Long and her journeyman were the only electricians on that job.

H E WAS AN OLD-TIMER, 61 at the time. Every day was like a countdown until his 62nd birthday when he was going to retire.

When something had to be done, he'd send me to go get a locknut or two yellow Scotchloks or something like that. I'd never get to see him working, he's always sending me to get material or go get coffee. So I caught on to him. My pockets were bulging full of locknuts, red Scotchloks, gray Scotchloks. Anything that I could think of that he would say that he forgot and needed to finish those jobs. He got wise to me. It was this unspoken tension back and forth.

He was the kind of guy who would put his back to you while he spliced out, so you couldn't see how to do it. I figured out early on, he's not ever going to show me anything. So what I did is, I tried to figure out, what was he interested in? He was into model trains. So I'd ask him questions about model trains. He'd be happy and talking about his model trains, then I'd throw in a question. "People keep talking about galvanized pipe—what is it?" He would answer it 'cause he was already in a good mood. I would get information out of him that way, in conversation but not by demonstration.

He would have these little tirades, screaming and yelling and carrying on. He would get pissed off because the coffee wasn't the way he wanted it. He'd say, "I should throw this coffee in your face. I wanted it X, Y, and Z and it's not." And he would also tell me that I would have to go and get his lunch. My attitude is, Hey, your legs ain't broke, you go get your lunch. Lunch is my time. I would have to go out to the store—which I wasn't planning to do, which I didn't want to do. But he came from the old school where, If I tell you to carry my tools, you carry my tools. To him I was an uppity woman, I should have just done what he told me to do. It wears you down to have to deal with it day after day.

Once it sank into him that I was going to try, his attitude gradually changed. He decided that I was going to be his proof to the union that he was a good electrician, 'cause he was going to make me The Best Electrician.

I was supposed to learn from him. Getting to know him was to my advantage. I remember one time saying to him, "You want to go out for a drink after work?" He was in shock. So I bought him a beer, we're sitting there and we're drinking. There were other construction workers from a big high-rise job that was down the street. One of the guys says to him, "What does your wife think about you working with a girl?" He said, "She doesn't know."

I think we were five or six months working together. I'm looking at him like, How could she not know? Then the guy says, "Didn't you tell her?"

So he says, "No, I told her that I work with a guy named Frank."

I said, "I'm Frank? Couldn't you think of a better name?" I didn't have to razz him about it 'cause this other guy was doing it. It opened my eyes.

For her next job Cynthia was sent to a radio station under construction, a large site with thirty or forty electricians where "half the morning was getting the coffee order and then I'd get two and a half hours actual working. That was a good experience for me, too."

THAT'S WHEN I STARTED TO see this "protecting journeymen's work." The guy that the radio station had hired as a project manager said to the foreman, "Whoever wants to solder, I want to see their work." And he says, "No, this is not acceptable. . . . That is not acceptable. . . . I'll take the ones who did these three." He didn't know who did them but I was one, because my work was clean and it was neat. Some of the guys, their work was sloppy or they did cold solders. I just happened to be lucky. But there was a lot of bitter complaining from the journeymen that the first-year shit apprentice should do a journeyman's work. The foreman just said, "Look, this is what the customer rep wants, this is what he's going to get."

I worked with a lot of different partners on that job. I did have the feeling that they were testing me. It was also the first time I encountered somebody who absolutely refused to work with me. Who would not even talk to me, would not say hello or anything.

I overheard a conversation (because the foreman and the journeymen talk to each other like you're not alive, like you're a piece of wood). One of the journeymen said, "Put Cynthia with so-and-so, because he's cutting in panels. She ought to learn how to do that."

The foreman says, "No, I can't do that. That would be a disaster." This was the clue to me that there was a problem. So I brought his coffee to him last one day and said, "Why won't you work with me?"

Basically, he says, I don't think women belong in the industry. And since women are coming in, I'm going to retire early. He would rather leave the industry than to have to work with a woman.

My only reason for wanting to work with him was because he was obviously a skilled electrician and knowledgeable. I wanted to work with the guys who I could learn something from and he was one of them. My way around that was to just go and look at the work after he was done, to see how it's done.

In the early years, the struggle was to get people to teach me, and some of the lack of willingness to teach me—and this is from conversation with them—is the attitude, Why should I knock myself out to teach you the business? You're going to get married and you're going to be home having babies. You feel like a junior psychologist trying to figure out what's really going on so you could combat that and get the learning.

Even within a given program, the quality and content of training for two apprentices could vary enormously, depending on the kinds of jobs they worked and, once they reached the jobsite, the journeyman to whom each was assigned.

Marge Wood took the plumbers union test in late 1975, during a flurry of affirmative action hearings in Madison, Wisconsin, that led to local efforts. In 1977, she was indentured and sent to her first job, a large university hospital. She spent most of her five-year apprenticeship on federally funded jobs and worked primarily for out-of-town contractors. Her male counterparts tended to work for local companies, where there was not only a wider variety of work but more chance for long-term employment—and therefore company investment in training.

I DID HEAR THAT THE COMPANY basically said, Give me the woman or minority who has been sitting on the list the longest. That was the criteria. There were over a hundred tradespeople on that job then and I was the first woman hired out there.

The first actual job I was doing was cleaning dirty sump pumps in a tunnel area that was really, really hot, and hauling stuff around. I was put with this crippled-up old guy who had been grandfathered. He used to work for a hardware store. When the plumbers license came in, they gave it to people who had that kind of job. He had no respect of any of the journeymen. He was nice, though.

During the course of that job I was also put on setting fixtures—sinks and toilets and things like that. I got a lot of hassle from that because that was considered the gravy work, the easy stuff to do, and yet I spent months on it. *That* partner they called Mr. Clean. He liked to say he had five wives and all the divorces were his fault, wear red bikini underwear and show me he was wearing it that day. I had my partner once lying on the floor going, "Come on, Margie, go for a ride on me."

There was, oh, a lot of sexual harassment. But I really never protested. What's the point? I actually felt that it was in a way *my* duty to get along with them, more than theirs to get along with me.

By and large I worked on big government-funded projects, one after another. It was a year and a half before anybody let me solder pipe. I did a lot of hangers and things like that. I have zero hours of residential construction or even light commercial. So I know glass piping, but in my apprenticeship I never worked with plastic pipes—plastic pipe is like everything. I worked only with cast iron because that's what's used in big buildings.

Early on, when I would find someone who would give feedback, I would want it too much from them. Then when they would say something like, Why do you always do it the hard way? I'd be crushed. The guy who said, Get on me and ride me, once told me that people were either born mechanical or not born mechanical. He could tell that I wasn't born mechanical. I believed him. I can remember riding the bus home fighting back tears going, Oh, I'm faking myself out, I'm faking the world out, I think I'm mechanical and I wasn't born with it. Universally, apprentices do not get the feedback to know whether they're doing well—not just myself, but other people, too.

Maura Russell knew she wanted to learn residential plumbing. She concluded, though, that that aspect of the trade could best be learned through a third avenue of training, rarely offered to female apprentices: side work. This was extra work on the weekend or evenings with a journeyman who was remodeling a bathroom, adding on a porch, or changing a service in someone's home. "Some of the men who were more experienced would have one of the apprentices come work with them on the weekend. They would learn something about the work that was different than in the union. I had *no* avenue to do that. If you exclusively learn on the job, you would not know that much about residential work. You just wouldn't."

To succeed, tradeswomen had to be flexible enough to adapt to the culture and organization of the workplace and maneuver through a system where they held little power. They had to make compromises, go along to get along, and sometimes this backfired. Paulette Jourdan started her plumbing apprenticeship with a journeyman who called her Josephine. Her last job before graduation was at a new jail, on "an all-black crew." Her foreman was someone she'd dated a few years earlier.

H E WAS A HELL OF a teacher, just—his methods were harsh. Because he wanted me to get it. I was about to graduate and we were working on underground sewer lines, and I hadn't done too many rolled offsets. Instead of, you offset to a 45 and you come back—this is up in the air, rolled. Now you've got to do some different figuring. He kept harping on it until I started to get it.

He never quite came around to saying, "Sleep with me," but it was in everything he did and said. There was so many nuances and so many subtleties and so much wisecracking and fooling around. I was riding there with him and three other guys, and all the way there and all the way back, it was just all these innuendoes. He was really starting to put a lot of pressure on me and I really didn't want to be bothered. But he was my foreman and I didn't want to have a rotten relationship with him at work. Because I wanted him to train me. I needed to get this piece—you have some nervousness about things you've never actually quite physically done. Like, I'd never actually laid underground pipe and piped it to different places.

I would just go home at night and cry sometimes. I didn't know quite what to do. And I think that's where I really regretted compromising myself, way down, because I let them talk all that shit around me and I never protested really. Up to a point.

I mean, he only went so far. The farthest he went was when I was down in a four-foot trench, trying to put a piece of pipe together, get the coupling straight or something. It was being difficult, and he jumps down in there with me—and my head is turned—and starts talking all this dirty talk. And I look around and he's got his thing in his hand. I jumped—I scrambled out of that trench. I walked away, I don't know where I went. I was gone for about an hour. You know, you expect your brothers to help at least. He was trying to help me, but he had this other agenda. He couldn't let go of it.

I just wanted to keep the peace and get trained and I didn't know quite how to handle it. My girlfriend was trying to tell me to set him straight. But I didn't have the courage. I thought I needed that last piece to call this training complete. I wanted to be able to say, I can do this, I can actually do this. And I thought, Anything I say to this guy is going to jeopardize that. Because he was running things. He would have laid me off, I think, if I had just bucked him in any kind of way.

I didn't know who I could trust, either, because it was him and me in that trench. Nobody saw that. It would have been his word against mine and my credibility was already very low in that area. I mean, enough people had seen us laughing and talking about fooling around—not really fooling around, because I'd get out of the way in time. But I never stopped them.

And I had just moved. My rent doubled. So when I started this job, I thought, I've got to keep this job. I've got $850 a month to pay in rent. So that was another factor that was getting in the way of my even gathering the courage to do something about it.

Apprentices entered training programs at a wide range of skill levels. At the end of their indenture, they graduated as journeylevel mechanics at the full rate of pay, competing for jobs with the co-workers who had trained them. Lorraine Bertosa felt she had to play catch-up through her carpentry apprenticeship in Cleveland.

I WAS 29 AND FEISTY STILL. Wanting to prove to the world. Probably fit right in with the guys. I don't know why I went in with such a passion, 'cause it was real difficult. My family was still struggling with alcoholism. I hadn't gotten help still for my insides. The biggest part was, I had to look good. I thought I had to look like I knew what I was doing.

But I'd never done hammering. Some of these guys had been with their dads for years and years. I still remember in apprenticeship class we learned to build steps. We had these little groups around the room. Me and this other guy, we were proud as punch of these three little porch steps. I had two guys in the corner building a winding staircase! I mean, fuck porch steps! Many of them had that hands-on exposure.

Because of the alcoholism in our family there was a lot of secrecy and pretending things were okay. When I started carpentry, I didn't know what I was doing, but I had to look like I did. With the additional piece of being a woman.

They look at you as an apprentice, period. But as a woman apprentice, people even looked at you more. If I'd go and pick up a piece of plywood, that'd usually quell their fears. I'd work. Every new job you'd have to show that. There's so much doubt about what a woman can do. It took a long time to be able to laugh at myself and say, "Hey, Joe, show me how to do this." I didn't spend my first four years like that. I think I'd have learned a lot more had I been able to do that. I've run into guys that haven't seen me in years, that remember me when I was the apprentice. "We couldn't tell you nothing, Lorraine. Couldn't tell you a damn thing, you were so defensive."

I didn't know what the hell I was stepping into. I thought, I'd like to build houses. That was some image I'd always had. It took nine years before I actually built a house in the union. They put me on bridge building. Then concrete form work. I wasn't crazy about doing concrete form work. It was big and awkward and very little detail. Once I got into doing trim on buildings it was a very specific size of wood. But here on the concrete it was, "I just need a block about so big, Lorraine," and they'd put their hands up and that was it. You'd cut it—"Oh, that's good." I had to sneak my way, push my way into doing different things. I got a job putting up metal studs in drywall. I didn't know shit about it, it was like, "Ha, I can do it."

I had to wield my way around. They came to my one crew, I was working form work. God, we were in this sewage plant and it was awful. The crew happened to be mostly housebuilders. They came to this crew because they were so skilled to say, We got this building over here that needs trimming out. Do you want to go on that job? It was this huge building complex, it had crown molding. The whole crew wanted to go and I said, "I want to go, too." They said, "Okay." Another woman was on that jobsite. I said, "All you gotta do is ask them." She said, "I can't ask." I said, "That's the only way you're gonna get on and learn anything else."

The plumbers used to have this deal where every six months you go to a new contractor—maybe somebody was doing residential, somebody was doing commercial, somebody was doing factory, so you'd get some broad sense of it. The carpenters—you could get left in the bridges forever.

There is an intimacy to any partnering in the trades, whether between two journeylevel mechanics or a journeyperson and an apprentice. Not only are they side by side eight hours a day, month after month, but they need to accommodate each other's work styles and habits, and often determine each other's job security and safety. It's not unusual for men partnered together to call each other "honey" or "dear." Likewise, partnering can be a clash of wills and prejudices. Lorraine spent over half her apprenticeship with one journeyman. Being a capable mechanic did not necessarily translate into being a skilled or willing teacher, particularly when the apprentice was a woman.

I'D GET AWAY FROM HIM, and I'd be back on another job—same guy! Sometimes even with different outfits. He didn't know how to handle it.

I would say that if a guy had sisters, and/or went out to college or went out in the world before he came into the trade, he got along with the women a lot better. If a guy just came out of his home, knew his mother, and got married, he sure as hell didn't know what to do with a woman on the job. He had no reference to go from, to seeing a woman other than his mother, who was probably at home cooking. R— hadn't had any of that. R— had been in Nam. Marine sergeant to boot. Back in the bush. Before anybody even went to Nam, he was in Nam. When we were alone he was real nice. We'd talk. He'd talk about Nam. He'd show me things in the carpentry. We'd do all this work. I know he liked me. But any time we were around other guys he was like, Come over here! and Shut up!—just talking nasty all the time.

At some point I remember going into my foreman's office. I threw my hat on the table and I said, "I can't work with this guy anymore." It was toward the end of my apprenticeship. I thought, God, I've got to start figuring out these projects. I still wasn't feeling that confident, but I figured I better practice. I'd start saying, "How about we do it this way?" "Shut up! I'm thinking!" I felt my brain was shriveling up in my head. My boss talked to him, he was good for a week, then it was back to the same old thing. Years after that, I saw how much I had learned from this guy. But I had to get away from him to pull it out of me. He just didn't know how to teach and he didn't know what to do with a woman. He'd say, "If you're gonna learn, you're gonna learn the right way!" That's good, to have such incredible expectations for someone. But you don't have to tell them in a gruff voice. The thing is, I think that's the way *he* learned. And that's the way he taught.

I didn't quite realize the magnitude of what I took on. You can't just do it like, Oh, I'm going to put on an apron and do this job and after eight hours I'm done. It's physically exhausting. It's mentally exhausting. People do this when they're fresh out of high school. They don't have a child in tow. They maybe aren't involved in a relationship yet. All their money's going to their brand new car. I was just barely getting by with an old beater car and trying to juggle everything in my life. I don't know what it would have been like had I not had all that. It might have been much more of a piece of cake, if I was 20 years old. Maybe having taken math or shop in high school. I can't even imagine that.

Being a foremother is no fun. 'Cause you don't know what you're doing. You're just doing it because you know you have to, at that point. I didn't want to be a cook anymore. I wanted to be someone else.

CHAPTER FIVE

Marking Gender Boundaries: Porn, Piss, Power Tools

I don't worry about the ones who say things to me. That quiet person with that very controlled anger is the one I worry about. You can feel the anger, they don't have to voice it, you know it's there.

And those are sometimes the ones who try to be the nicest to you. You have to watch them.

—Gay Wilkinson, Boston

Close to eleven on a Friday morning, the steward was walking around the 44-story job collecting $2 each from the roughly sixty electricians on the site to celebrate the general foreman's fiftieth birthday with a drinking party in the shack. The party would start at lunchtime and extend into the afternoon. A stripper would be performing.

I was, at that point, less than a year out of my time.

Several of the new journeywomen in my local, including myself, and several of our business agents had only recently gone through a training together on sexual harassment. Earlier that week a highly publicized rape in the Boston area—on a poolroom table at Big Dan's Tavern—had called public attention to sexual violence. And it was the same week as International Women's Day. Ignoring the situation didn't feel like an option.

The steward told me that I didn't have to contribute or come to the party. I countered that, if the steward was organizing a celebration of the GF's birthday, it should be done so that everyone could participate. And I explained why I didn't think there should be a drinking party with a stripper on a union jobsite. "Just because we have to take you in," the steward said, "doesn't mean anything has to change because you're here."

I knew I didn't want to go to the party or be working on the job that afternoon. I told my foreman I was going home. Before leaving, I called the union hall and told my business agent that I was walking off the job and why. He asked what the other two female electri-

cians there thought. I said that since both were apprentices and more vulnerable, I hadn't talked with them. He explained that, given how late it was, there wasn't really anything he could do. I said I understood. And I went home. Expecting the party to go on.

Monday morning on the bus ride to work, I learned from a woman plumber who worked on the site that, after I'd left on Friday, my business agent had asked the steward to cancel the party and return everyone's money. My breath caught. I was surprised and impressed that the hall had acted, but I knew there would be retribution.

—Susan

On jobsites the behavior of those in authority—the foreman or general foreman representing the contractor (though they are also union members) and the steward representing the union—set a tone and an example for the crew to follow, and strongly affected a tradeswoman's sense of her welcome and safety. On her first job as an apprentice carpenter Lorraine Bertosa felt protected.

I REMEMBER MY FIRST FOREMAN LITERALLY saying to the guys, "Watch how you talk." He said that in the first week I was on the jobsite. He was one of these guys that felt confident himself, wasn't out to prove anything. It was fine that women were there. A really unbelievable guy to get as a first foreman. If you were willing, then he was willing to meet you halfway. He would say to the guys, "Don't talk like that. You can't talk like that around here" (cuss words, certain things they were saying). I think that pressure came directly from the office, from the contractor. We want to keep these women.

Where contractors and unions did not make such a clear commitment to "keep these women," new tradeswomen were less fortunate. Co-workers, foremen, or stewards who felt that women did not belong in the industry at times expressed that opinion through words, actions, or silence. Before affirmative action brought government support for a more diverse workforce, harassment, ranging from petty to criminal, had been a standard means to discourage those who strayed across the industry's gender and racial boundaries. It did not end when the government regulations began.

Tradeswomen were sharp observers, and most perceived themselves to be on their own in handling any hostility. They worried that requesting assistance could as likely bring retribution as help. Given the imbalance of power, many women put blinders on, kept their focus on the day's work, and waited for a bad situation to end by itself. Women, especially those unfamiliar with the safety practices of tools and equipment, were particularly vulnerable on their first jobs. Not only were they green, but they were not yet sworn into union membership. Probationary periods could range from a few months to two years, for those entering under special affirmative action guidelines. Kathy Walsh was sent driving on a wild goose chase looking for the foreman on her first day at work—hazing that might have happened to any new apprentice. But on her second day, when she knew where she was going, the ironworker who'd verbally expressed his resentment about having a woman on the job expressed those feelings again, this time physically.

E VERYBODY PARKED UP ON TOP of this embankment. It was about forty feet down to where we were working, very steep, and it was muddy and slippery. An ironworker pushed me from behind. And I slid most of the way down that embankment face first.

Getting up from there—I can't remember whether I was crying or not, if I wasn't I was almost—and getting the mud off of my face and out of my tool pouch and going to work that day was one of the hardest things I'd ever done at that point. Mark, the guy that was nice to me, was like *so nice* to me that day. He gave the guy shit about it, and he came down as quick as he could and helped me get up. At the end of the day he said, "I don't know anybody that wouldn't have walked away at that point. You just keep it up, and fuck these guys." My first day I slammed my hand in the car door. My second day I went down face first down a muddy forty-foot embankment.

The job lasted for about two weeks. They laid me off and I was like—*uh*. I think I made it back to my car before I started crying.

Loyalty by trade is very strong in construction. Workers generally spend coffee breaks and lunch: carpenters with carpenters, ironworkers with ironworkers, painters with painters. For a journeyman of one trade to push down an apprentice of another trade is highly unusual, because normally, the full crew would rally to defend *their* apprentice. Attacks on women put men in the position of choosing between male bonding and union or trade solidarity. Only one of the carpenters came to Kathy's assistance. When she reported back to her apprenticeship coordinator after the layoff, she never mentioned the ironworker's action, or the tacit approval of most of her crew. "I was totally intimidated by the whole process, all of it. We didn't even join the union until we had at least 600 hours in."

The behavior of the union representatives a tradeswoman happened to encounter was critical to shaping her expectations of whether or not the union would assist her in handling harassment or discrimination. Although MaryAnn Cloherty would return to union construction years later and complete her apprenticeship with a different local, she quit the first time around. She was a second-year apprentice on a job where having a steward on the site only added to her problems.

THERE WAS A LOT OF pornography on the job, and when I would complain about it they would take it down and they would put up more. Crotch shots, legs spread, blown up. I mean there was a crotch shot that was blown up that was at least three feet by five feet. I walked by it for three days, I didn't know what it was. I did not know what it was until I was on the other side of the picture and I saw a whole series of porno shots. I realized what the other shot must be. That was when I complained.

The offending stuff came down. And then the next day the whole jobsite was littered with it.

There was a union steward who was the worst offender. I really felt like there was nowhere to go. My steward when I first arrived on the job said, "Put your tools over here." After I put my tools down he said, "One thing you got to understand is, I used to throw gooks from helicopters in Vietnam." I didn't know what was that supposed to mean to me. I think he was trying to scare me or intimidate me or paint himself as a big ogre. I didn't really think I could relate to this guy.

A skilled construction worker must be able to climb scaffolding, use power tools, lift heavy objects, and perform countless other tasks that are inherently dangerous. But like driving a car on a freeway, they can be accomplished with relative safety given proper training, support, and equipment. Just as a student driver wouldn't feel comfortable in high-speed traffic accompanied by a driving instructor who was threatening, someone learning to splice live wires, walk an I-beam, or maneuver their way through the obstacle course of a construction site needed to trust their supervision in order to focus on the actual task at hand.

As a first-year apprentice plumber in Boston, Maura Russell was sent to a new building under construction, a good opportunity to see a project from the ground up. On the crew, though,

ONE GUY WAS REALLY A very sick fella. One day we were both carrying a length of 6- or 8-inch cast iron pipe. It was a stage of the underground, and he was on one end and I was on the other. We were carrying it from one place to a trench on another part of the job. We were walking by this one big pit that had all this rebar, reinforcing bar, sticking up in various patterns because they were going to be pouring a floor and also have some starts for some columns.

He gave me a shove with that pipe so that I went down into that pit with the pipe—which is heavy pipe. And it was really lucky—luck had a lot to do with it—that I landed on my feet, still holding the pipe. That I did not end up in a perforated sandwich, with the pipe on top of me, landing on a lot of that rebar which was vertical. I can still see him standing at the top of that pit with his little Carhartt jacket and reflector shades and Arctic CAT hat looking down. And with his little psycho voice saying, "Gotta watch out. You could get killed around here."

He was really creepy.

I'd be pouring lead in a pit, in a trench. It's a sunny day. This is totally outside. All of a sudden, cloud. And there'd be this Dick—which was his name, actually—totally bending over me, blocking the sun and whispering in my ear in his little creepy voice, "Watch out that you don't get any water in that lead. It could pop up and you'd get a face full of lead and that wouldn't be too pretty, would it?"

Rather than bring the danger she felt from this journeyman to the attention of any authority, Maura just dodged him as best she could. She recognized the box he had her in—it was her word against his. And what's wrong with his warning her to be careful? And who wouldn't believe that a green girl apprentice simply lost her balance carrying heavy pipe?

Women who had no reason to perceive the union as offering them protection, but were still committed to staying in the trade, often chose not to report even very serious harassment. Karen Pollak had applied to several Kansas City unions over the years before affirmative action regulations created an opening in the Carpenters. Having learned the trade from her grandfather, she passed the journeyman's test. She was allowed to enter as a first-year apprentice. Despite the opportunity to hire a skilled mechanic at apprentice rate, it was a year before a contractor would hire her. On her first day on the job as an apprentice carpenter, she could have reported her treatment to the union. Or to the police. Committed to keeping the job, she chose instead the silence she felt was required.

Since none of the carpenters wanted to work with her, Karen was partnered with a laborer who was "none too happy to be working with me. He was trying to do everything he could to drive me crazy. I lost him for several hours in the afternoon. I couldn't find him." Assigned to put in insulation at the edge of the building, Karen was given a safety belt that was too large for her. She eventually just left it "hooked up onto one of the lines, but it was laying over the edge of the floor." When the superintendent found her still working later that afternoon, he told her he'd assumed she'd fallen and died. While the super was admonishing her for not wearing the belt,

I LOOK DOWN AND THE LABORER that I was paired up to was taking a sledgehammer and just demolishing my little red Volkswagen. It was like, "What did I do?" Well, he explained to me that we don't drive Communist cars onto union parking lots.

I couldn't leave my tools at work, because the gang boxes were full. I'm over in the middle of nowhere, with no way to get home and I can't leave my tools. So I just put my toolbox on my shoulder and we hitched a ride. This farmer picked me up alongside of the road about a half mile from the jobsite. I got home, though, several hours later than I should have. And the husband was real upset. He was like, "Where's your car?" That was the nicest car that we owned. "Well, we don't have it anymore." "What do you mean, we don't have it anymore?" And then I explained. And it was like, "Well, you have to press charges against this guy. You can't let him do this shit to you." "No, I can't do that. You don't understand. I *will* get pushed off the building. You can't do those things."

I eventually got it towed home. We used parts off of it. I had nice seats and a nice shifter. But as far as the car—he had taken a cutting torch and cut the frame. I would assume it would have to be on work time, because I had the car at lunchtime. When I went back to work from lunch, it was fine.

After getting chewed out by this superintendent because I had left my safety belt and it was hanging over the edge of the floor and he thought that I had died, it was like, "Did you even go down to see if I was there?" "No, I just figured I'd worry about it when I got down there." Well, that told me where I stood. So that's why I was not going to press charges on my little Volkswagen. We just gritted our teeth and went on and bought a really old Volkswagen, and took and drove it to work. But from then on I parked it two or three blocks away from the jobsite.

These were real strong-valued people. It was not a union-made car and it represented to them, definitely I had to be a communist. I was driving a Volkswagen. I was a woman wanting to be a carpenter. So I had to be. That was my first day of work. Welcome to the real world.

Asking for help was not necessarily a more useful response, as Yvonne Valles learned. Attracted by the opportunity to work with her hands and the hope that she'd be able to buy a home once she made a journey level painter's salary, she was an eager first-year apprentice. She joined painters hanging vinyl wallpaper at a hotel in Los Angeles, and within the first two weeks faced harassment from her foreman.

I'M STILL KIND OF TRAUMATIZED by the second job I got. The foreman on the job was a real jerk. Him and a couple of the other painters would always be talking real dirty about women all the time. They used to leave magazines of naked women in the bathroom that I'd use. They'd leave the book wide open and it would show. They'd think it was funny. They were harassing from day one.

There was a young kid apprentice that was about 18 years old. My foreman used to talk verbally abusive to him, call him a dickhead and all kinds of names. With me, I heard him making a crack one time, called me a dyke. Anyway, he was always bragging on breaks. He'd be talking to the guys, but I could overhear him because we'd eat in the same room. I mean, where was I going to go eat lunch? He used to pick up prostitutes. He'd be saying, I'm going to see so-and-so tonight.

One day, I was hanging up some wallpaper and he came to me, I was kneeling down. He goes, "Hey, you want to see some pictures of my girlfriend?" And I said, "No." He said, "Oh, come on, I'm training her to be an apprentice, too. Don't you want to see some pictures of how I train my apprentices?" I said, "No, why don't you just leave me alone?"

So anyway, I was kneeling down, spreading the wallpaper on the walls. All of a sudden he stuck a Polaroid picture in front of my face and he goes, "Look." And I looked. And he starts laughing.

It was a picture of a young woman laying down with her legs open and she had what they call in wallpapering a seam roller. It's got a little handle with the roller on it, you lay the seams down flat with that to get the air bubbles out. She had the handle inside her vagina. And he starts, "Yeah, that's how I train my apprentices."

Oh, man. I just said, "Get out of here, I don't want to see that!" I was really upset. I went home that day and I called the apprenticeship school and I told the head of the apprentice school, "I got a problem on the job. I'm being harassed and I just want you to know what's going on."

I told him about it, and I started crying 'cause I was really humiliated. He says, "Oh, gee, I'm sorry," and "That asshole," and he goes, "Yvonne, it's not always going to be like that." He says, "I'll talk to him."

But nothing ever happened. He had told me too, "You know, Yvonne, I can report this but it might not be good for you."

I said, "Well, there's only one thing I'm afraid of. I've heard that women that file lawsuits against their companies, they end up getting blackballed. I wouldn't want to have that mark against me." He said, "That's true, that could happen."

They don't care. They want to discourage you. It's like contractors have this attitude, from what I've heard, if a woman sues them—fine, they won't hire any more women at all.

I hated that guy. He was disgusting. He used to ask me if I'd want to snort some cocaine with him after work. I just kept my mouth shut because I needed the job. I needed to pay my rent, so I just tolerated it.

Any new worker wants to make the workplace more comfortable by developing congenial relationships with co-workers. Yet as Melinda Hernandez, a new electrical apprentice learned, friendliness could set off an invisible minefield.

ON THAT JOB THERE WAS an apprentice—he wasn't a piece of shit, he was *the* piece of shit of life, the lowest of the low. But I didn't know this, see. He came off very nice. He happened to be Puerto Rican, too.

And he says, "Oh, it's nice to have a girl working side by side, why don't we hang out one day? We'll go out to dinner after work." So I didn't know. "It's just dinner. What's the big thing? What, are you afraid of me or something?" But he came off very nicely, so I said, All right. Maybe I can make a friend, you know, in the industry starting out.

So we went to dinner, and after dinner he wanted to go out dancing or whatever. And I said, "No, you know, I told you that I have someone, that I'm involved."

To make a long story short, that Monday we went to work, I think he told everybody what every man wants to hear—that we got intimate (and that's a very refined word coming from this character, okay). He did me, you know.

He became very nasty, openly, verbally cursing a lot, talking about who he screwed the night before to the men. And I'm sitting in the men's locker because the women weren't given their own locker. One day he actually brought in pornographic material, pictures that he had taken of a woman close up, with a flashlight. The reason I know this was because he was describing to them the pictures when I was in the room. And they were laughing. But none of them ever took a stand. I thought in their minds they figured, Well, it's not my daughter, or It's her own kind, it's a Puerto Rican just like her doing it to her. It's not us. Whatever it was, they justified it. Nobody ever said anything. And I remember there was a guy in the room that was sitting in the corner, he was a born-again Christian, reading a Bible.

I got up and I walked out, I just stepped out of the room. I realized that I was in for a long haul, because that was my first job. Wow, you know, what a drag. But I hoped. I had high hopes that things would get better.

Family support was key for Cheryl Camp when she faced hostility on her second job. The knowledge that her union rotated apprentices to a different shop every six months meant that even if treatment didn't improve, it would at least end. And the fact that men on her first job had been particularly supportive helped her ride through the hard times.

THERE WAS AN ELECTRICIAN ON the job, a younger guy, too. And a minority, he was black. He went out of his way to harass me. It really irritated him to know that there was a female electrician on the job. And plus, I was an apprentice. He had gone through the trainee program and, you know, there was a stigma always attached to the people that came through the trainee program. I can't repeat the things that he said. He had the filthiest mouth, I mean really filthy the things that he would say. And then he would describe his outings the night before with ladies of the night and go off into really intricate details of his endeavors and make sure that I could hear every single word. If I was walking someplace, he would start walking behind me and making rude comments about how women are.

What I really hated was, all the guys on the job knew that he was doing this to me, that he was harassing me. And no one intervened and talked to him to tell him, Why don't you back off and leave her alone. They knew that I was new, that I was an apprentice, and as an apprentice you're supposed to be seen and not heard, you're lower than whale crap. You really aren't supposed to have anything to say to a journeyman as an apprentice, other than asking questions, *if* they allow you to ask questions. I really don't think that they even considered my feelings in the matter. When I told them that I was taking him up on charges for harassment, they told me, Well, this is just the way he is, and Don't let it bother you. But that's impossible for it not to bother you.

There was another female on this job. She was a plumber, but we didn't work in the same area. He was harassing her too, but her husband is also a plumber, so he straightened him out so he didn't say anything else to her. But I had no one to intervene for me. And he was the type of individual that you could not just approach personally, and say, Why don't you just back off and leave me alone. It was the foreman that came through and ended up having him apologize to me. He ignored me after that.

I was under so much stress with him, from what he was saying and the way that he made me feel every day, I was ready to quit the trade at that time. My mother talked to me and was saying, "Well, Cheryl, you don't remember what your ultimate goals are. You wanted to finish this and see it through. You know the first shop that you worked for was so great and the guys were different there, so it's not going to always be this way. Just bear with it and try to see it through and it could get better." My mom was a real source of comfort.

The effect of a harasser's action was compounded when others on the job knew about it but did not intervene—as though he were acting on their behalf. Contractors and unions tended to underestimate the gravity of harassment and in some instances even condoned the behavior, tacitly or explicitly. Institutional procedures for prevention or punishment were rare.

Acts of passive aggression could cause serious injury without anyone's seeming to be responsible. Although with an inexperienced worker it might be difficult to distinguish between a true accident and an intended one, it was the responsibility of the supervising journeyman to look out for an apprentice's safety, and the responsibility of the training program to properly prepare apprentices. Karen Pollak saw the failure to train apprentices in the proper use of power tools not as some malevolent attack on women, but merely as the result of assuming that apprentices knew how to use them, which had traditionally been true. Karen had been trained to use a skill saw safely when she was five or six years old (by her grandfather, who showed the grandchildren his missing finger). But other female apprentices received

L OTS AND LOTS OF INJURIES. Eye injury. Feet. Hands. We had a woman that lost three fingers. Because no one told her how to use the table saw. Another one was cutting stakes out on the jobsite, cut off her whole hand. All because no one took the time to really, really explain that these things can hurt you. I knew how to use the tools. I had an unfair advantage to a lot of the women. Basically what they taught you was how to put the saw blade into the saw and make sure that the guard worked, if there was a guard. That was about it.

They would say to the woman that it happened to, See? I told you, you should have stayed home. A broom wouldn't do that to you. And then they would make it a point that *you knew* that someone had gotten hurt.

They told me when the lady cut her hand off. She was using a big radial arm saw, a 16-incher, out on the jobsite. She had put her hand down to hold the material. The material started to move. The saw got bound. And somehow or another her hand got back behind the saw, so it pulled itself right back across her. They were able to save it, but she didn't have full function of her hand. It's not the same. And never will be. That's something that could have been easily prevented.

Even if it began as an unintentional oversight, once women started to experience so many injuries, an adjustment should have been quickly made to incorporate power tool safety into the training. Instead, the pattern of accidents became not only proof that women didn't belong, but an amulet to frighten women into leaving.

Some job situations had the feel of trench warfare. Men who wanted to drive women out; women who were determined to stay. Knowledge of tools and experience at the trade did not prevent an "accident" that broke Karen Pollak's nose, when a journeyman did not want her—not only a woman, but a Cherokee Indian—working with him.

I HAD A SLEDGEHAMMER DROPPED ON me. This was a job that they had to have a woman. And they needed a minority. It was like, Give me a black woman or somebody who I can mark as a double and then I only have to have one of them. It was just a little tiny library for the University of Kansas Medical Center.

We were down in the hole and I was stripping forms. The guy above me was on the next set of scaffolding working on the next layer. I kept noticing that hairpins, which are a form-type hardware, would fall down and hit the hardhat. Every once in a while it'd hit the bill and knock the hat off. You'd bend over and pick the hat up, look up and go, "Can't you be careful?" "Yep, I just dropped it, sorry."

The superintendent had yelled at him about something. I was standing below and he was going, "Well, make *her* do it. She doesn't do anything."

"She's stripping. That's all she's here to do."

I was going, "Well, I'm willing to learn. I can handle doing more things than just pulling nails."

"Nah. Not with me you're not." At lunchtime, the foreman said that I was going to go help him after lunch.

He got up on top of the wall before I did. He was standing up on the scaffolding he had just built. I was just starting to climb up the form. BAM! The sledgehammer hit me, it rang my bell.

It was like, Okay, that *could* have been an accident. He throws the rope down. I hook up the sledgehammer and he pulls it back up. I make sure I'm away from the rope. If he happens to slip again, no problem.

For some odd reason, he didn't nail down his scaffolding like you're supposed to do. He told me he did. I stepped on the far end of the board and the scaffolding went smack with the board right in the face. Straight down, back into the hole.

The hole had mud in it. And water. I had hip waders on earlier that day stripping it out. It was an ugly sight. I had broke my nose. The superintendent comes over and says, "Well, this isn't going to work. He doesn't really want you up there."

"Oh, I just thought it was an accident that the sledgehammer fell."

"Probably was on his part, Karen."

"And that's why he didn't nail down the boards, huh?"

"Well, maybe he was getting ready to move them over to the next set of scaffolding."

"Right. He knew I was climbing up there."

I stayed on the bottom and stripped. He would drop things if I was underneath him. I soon got the idea, Stay away from him.

The wisdom of Karen's response—to outwit her journeyman's efforts to injure her while keeping up production—is made clear by the actions of the superintendent who both represents the contractor on the job and belongs to the union. The journeyman responsible for her safety not only drops a sledgehammer and other objects on her, but lies to her about the scaffolding being nailed down, resulting in her fall and broken nose. Rather than laying off the journeyman or bringing him up on charges in the union, the super accommodates his wishes. All three understand the same unspoken ground rules: not only is it acceptable to refuse to work with a woman, it is acceptable to communicate that refusal through actions which, out on the street, could result in prosecution for assault and battery with a dangerous weapon.

Harassment could result not only in a stressful work environment or physical injury but also in economic costs, both short- and long-term. It was not unusual for a tradeswomen to be transferred or laid off after attention was called to harassment. Barbara Trees found that her skill training was also affected.

I WAS A SECOND-YEAR APPRENTICE working for this contractor doing ceilings—the concealed kind, the hard kind of ceiling—and I was really trying to learn them. The bar isn't revealed, you don't see it, so they're kind of complicated. I wasn't finding it easy to begin with. I was up on the Baker [staging] by myself and the electricians opened up the computer floor around me. They opened up enough tiles so I couldn't move my Baker. I said to them, "You know, I need to move this Baker. Will you put back those tiles?" They just wouldn't do it.

I'd be working on a Baker and they'd be having their coffee break and I would hear my name fairly continuously. "Barbara ... Barbara ... Barbara ..." I got sick of it, so I called over to them and I said, "Is there something you want to say to me?" And, "Oh, no, no, there's nothing we want to say to you."

That was really all it took.

I went home that night and I came back into work the next morning and these guys obviously had written on my Baker in letters a foot high, "PROPERTY OF THE CUNT." I didn't know what to do about it. I didn't really think there was anything I *could* do about it. But what happened is that my sub-foreman came over and he saw it. I didn't really want him to see it or anything. I was embarrassed, actually. So he says, "What's this?" I said, "Well, I think those electrician guys wrote this on here, you know, because we had words yesterday." He says, "Well, we can't have this. I'll speak to the foreman." I was really surprised by his reaction. I felt he was trying to help me.

So the foreman came over to me and asked me what happened. And he said, "We can't have other trades harassing our carpenters. We're going to have a meeting of all the trades later in the day. I'll let them know that they can't do this." I thought, you know, this sounded good, this sounded like what he was supposed to say to me.

I guess they had the meeting and what happened is that I was transferred out of there. They just decided that I was too much trouble. I remember as I walked down the street I passed one of those electrician guys and he almost tried to hide in the building. He saw me and he kind of put his eyes down. I think he must have realized what he had done.

This was my introduction to how they help you out. This guy was so sincere—Oh, we can't have them harassing our carpenters, we're not going to put up with this. I remember thanking him, thinking, This is really great.

Whenever I see ceilings now, I sort of panic. I actually kind of get a cold sweat going. I started to realize what it's from is that the two opportunities I had to learn ceilings both ended where I was laid off or taken off the job. I still feel bad about this. That was my chance to learn.

So this is what happens. Your training suffers. You feel guilty. You don't know what you did wrong. You're feeling like kind of an awful person. And you don't learn your ceilings.

A tradeswoman who changed contractors or crews—particularly if she bounced between situations where co-workers were friendly and fair and situations where she faced hostility and humiliation—could find it hard to build self-confidence and gain her bearings as a developing mechanic. Like many tradeswomen, Helen Vozenilek, an apprentice electrician in Albuquerque, struggled to understand the cause of harassment, looking for how she might prevent or avoid it.

I MADE THE SHITHOUSE WALLS. IT was something like, FUCKING LESBIAN ELECTRICIAN BITCH. I somehow knew that had to be me. You know, process of elimination.

On that job, the steward was terrible. I don't think he liked women. It was just a bad collection of people. You know how men can get—when they're alone, they're fine, they're actually brothers? And then they get in a group and they're just beasts? I think that was the situation there. They sort of got beastly.

I did feel really harassed there and I didn't quite get it. I remember going home a couple of nights and just crying myself to sleep. I think the steward had talked to me that day, said they were going to run me off or something— and it was like, I didn't get it. The hardest thing is the capriciousness, not knowing what you did or what you were being held responsible for, or irresponsible for.

Some of the men who saw tradeswomen as invaders of their domain marked territory with graffiti, pornography, or bodily fluids. Although Irene Soloway, a New York City carpenter, "really didn't experience a great deal of sexual harassment,"

I HAD ONE INCIDENT THAT UPSET me for quite a while, which was a job that I was determined to do well on and keep. After six months of coming in every day, I took a day off. When I came back, the shanty had porno pictures all over it, real disgusting ones.

The foreman and I didn't get along. It turned out his brother had spent the entire day plastering the shanty on company time. I really was truly shocked, because I had been on the job for six months and pornography was not an issue. So I felt it was terribly personal. It's like, you don't even know where to look! The men were all sitting on their benches and I knew that they weren't comfortable with it, either. I mean, you have a shanty that's clean and decent, you have your little nail and your little hook and your little lunch, and then one day you come in and it's—you know, open cunts all over the wall. It made a lot of people uncomfortable, but I knew that nobody would say a word. I had a screaming fight with the foreman outside the shanty. I was a second-year apprentice.

The reason why I had a problem with the foreman in the first place was because I told him, in front of a group of men, "I'm an apprentice, I'm not an animal, and if you want to call me, I have a name." He used to call black people Nigger, you know. I guess he must have called me Girl. He was furious with me and he told me so in no uncertain terms. And then this happened. I ended up being sent off the job, and I never felt that I had any retribution for that.

They sent me to another job and then two days later I got laid off. So they sort of diffused it that way. I always felt I had to find some like really, really remarkable way to turn the situation around. You think you have to deal with this on your own and you have to be able to stay in the industry, you know. That was my philosophy. I always thought of how would I turn this around and have him be shocked and upset and angry. Which is stupid. I mean, he's the boss. I'm not. So you can't turn it around in that way, in a personal way. But that was the way I used to think.

As an apprentice plumber, Maura Russell never had the opportunity to work with another woman in her trade. On one job with several hundred workers, though, she was able to work with two other tradeswomen, an electrician and a taper, building apartments for the elderly.

W E HUNG AROUND TOGETHER, WHICH was really nice. But they have a hard time with women getting together on the jobs.

There was one time when the three of us were having lunch in K — 's car. This one guy who was there just for two days, an asphalt contractor putting in the parking lot, came over to where we were eating our lunch and pissed on the side of the car. Quite unbelievable. Looking at him coming over, at first I thought his truck must be parked next to us. And then K — is like, "Wait a minute! Is he doing what I think he's doing?" Really.

He'd left by this point. We convinced K — that what she should do is talk to the super. The guy's in his truck and he denies it. K — says, "Oh, bullshit, you did this." At which point, the super went insane that she swore. And just said, "If you talk like that, you don't deserve to be treated like a lady."

He was just going to walk away and not deal with it. And that was the point at which K — took out her little Swiss Army knife and told the guy that she would slash his tires if he didn't apologize. He'd been laughing at the incident, which is what really enraged her. He finally did say, "Oh, I'm sorry, I'm sorry, I'm sorry."

She said, "No. You got out of the car to piss on my car, and now you have to get out to apologize."

So he did. He was angry at that point that he was compelled to do that. And the super was jumping up and down livid, like she was a maniac. He didn't want her on his job anymore. He went running to the trailer to call her company to get her fired (which he was unsuccessful in doing).

But that incident later, we joked, would become that she had a ten-inch knife or something to this guy's throat—and it has pretty much gone around the circuit like that. But that was definitely, Talk like a lady if you expect to be treated like one.

Laborer boss said that to me later in the day, "My wife, she just would have turned her head."

Really.

Although, for women, responding in kind to harassment could bring on more trouble, Maura was sure that if there had been three men in that car,

THEY WOULD ABSOLUTELY HAVE PUMMELED him. And what would the super have done about that? He would have turned his little head. He would have expected that. Oh, they would have gone insane if something like that had happened to a man.

That was really quite an interesting statement of, You're not welcome. Gross, really gross.

Hostility could be triggered by small acts of self-empowerment. Like three women sitting together in a car. Or a woman becoming more assertive. Doubly vulnerable, as an African American woman, Gloria Flowers found that her worst harassment came when she decided to speak up for herself, after she reached "a point where I wanted to have some respect, I wanted to be talked to like I had some sense."

TOWARDS THE END OF MY apprenticeship I was really catching the blues. That last year, I said to myself, I'm not taking this crap anymore. I'm going to start telling some of these guys off. Well, that was the worst thing I could have done. It's almost like, when you get revenge, it's not as sweet as you think it's going to be.

That last year was my worst year by far. I remember this super telling me, maybe I shouldn't have gotten in the trade. "Why don't you just give up and give out?"

I fell out with a lot of the guys. Some of the guys I had liked previously, we ended up just rubbing each other the wrong way. They started rotating me, working me every other week. At the time I didn't know it was because they wanted to lay me off and couldn't figure out a way to really do it.

This one job I was on, the Ohio Bell Building, downtown Cleveland, there was this black guy on the job, he hated my guts for some reason. He had problems with women, he was like in his third or fourth marriage. That was the worst, the darkest period, I have to say, because he got physically abusive.

He pushed me, physically pushed me down stairs when nobody was watching, in a sub-basement. I remember being so mad and so hurt, I wanted to kill that guy. But he was a body builder, he was really built.

A lot of times they had raffles for different things on the jobs. It just so happened that on this job—I don't know why this happened to me, God was trying to show me something—they were raffling a .357 magnum. I don't know what made this guy ask me if I was interested that particular day. Generally I had my little blinders on. I was kind of kept in the dark on a lot of things.

But that particular time—and feeling the way I was feeling—I wanted in on that raffle. I remember coming home. I talked to a girlfriend about it, and she said, Pray about it, and don't do nothing stupid. You can't take on no man, blah, blah, blah.

She really brought me back down to the ground. I prayed about it and it ended up working out. He got laid off, and I got laid off shortly thereafter.

That guy wanted to hurt me. He did. I never told any of the guys about it because, you know, they didn't care. That job had gone sour for me. None of the guys wanted to have anything to do with me on that particular job because I just wasn't taking any stuff.

Accidents set up against women or people of color were particularly insidious. When "successful" they accomplished two things: eliminating or frightening the target, and framing them to look incompetent, not only as individuals but, by extension, as a representative of their gender or race. As an apprentice electrician, Nancy Mason learned to be extra cautious, in case work was sabotaged.

I WAS DELIBERATELY SET UP, ACTUALLY, on two occasions. Once I had circuits turned on when I was trimming out receptacles in a high-rise office space. I don't know who turned the circuits on. And another time I'd been hooking up fire alarm exit signs and I went back to check some. I was up at a light exit sign, and someone had actually tied the ground wire into the hot wire on the other end of the Scotchlok and as I was taking it out, someone turned it on and it blew up in front of me.

I did not get hurt, but obviously someone was hoping I probably would have, or gotten scared or whatever. Those incidents both happened, I think, when I was a fourth-year apprentice. I was getting pretty tuned in to always checking stuff with my own meter. But the turning on the circuit while I was at that exit sign was probably the most dangerous thing, because of the higher voltage. It was a 277 situation.

Bernadette Gross, who went through her carpentry apprenticeship in Seattle, was on many jobs where "the object was to buck me off, and I rode them like that. It was like, I'm not going anywhere." But harassment, even when handled, carried an additional personal cost—to one's sense of trust in other people. On a job early in her apprenticeship, Bernadette fell from a ladder that was not properly secured.

I WAS UP ON A SECOND story framing a window, and the ladder wasn't tied off and it slid from under me.

I had a sheet of plywood dropped on my hardhat. I mean, it hit my hardhat really hard. I was bent over and it could have broke my back, really—and there was just never anything done.

At that time I didn't have sense enough to think that someone had set out to hurt me. Just later on, it was like putting it all together. I was still pretty new, right? And then, I never had that many accidents after that. In my second year, I knew better. If somebody told me to go up a ladder, I'd check it, you know. But in life, it took me a long time to believe that people had malice in their heart. I always believed that they were sort of going their way and you got in the way and they knocked you over. It wasn't anything that they set out to say, I'm going to knock her over, you know.

But I found out that there were people who did, you know, sit down and plot that. It's kind of a hard blow for me.

Such experiences happened to women who graduated from apprenticeship programs. They cast an ironic light on the common explanation for those who did not, the new women apprentices who quit after only a day, a month, or a year, the ones who supposedly "found it wasn't for them."

(and yet) Passions

from PARTNER #6

Brought their homeselves in. Got
comfortable, work and talk flowing like

wire through well-laid pipe, or the tunes
they sometimes sang. One by one

the crew found cause to buzz close by
like homesick travelers drawn to warmth.
— *Susan Eisenberg*

— Barbara H. —

Not knowing anything. And then! once you learn ...

I can go into this house here and open up a box and actually look in there and get an understanding of where the power came from and what switch is doing what, what wire is doing what. And that still amazes me.

I like to look at a set of blueprints and know exactly what's going on, you know, where it all started and then where it ends up, the finished project. I get a high off of it just unbelievable.

— Lorraine —

There's always been this sense of total interest in big things. My brother is a pipefitter for the railroad—he's in California now—but when he was in Cleveland I went down to his shop a couple times. They'd have these trains inside this building, these locomotives. They'd be running. They'd be inside this building that's the size of God-knows-what to put four trains inside—Paul would take me up inside and pull the throttle. The power and the size is just, *Wow!*

Ironworker Mary Michels working on the Citicorp Building in Los Angeles (about 1985).

I worked on the Sohio Building for three years, I was in heaven. The vibrations and excitement of these people actually putting this together, it was ongoing. Everything was cooking.

It's like being a kid. We get to play with this and fix it and put it together and make it go. Make it be *actually there*. There's nothing—and we put something there, and it's there, it's like a rock. It stays, even. I mean, look at this amazing building that goes up to the sky!

Part of working on the Sohio Building that I loved a lot was being around the ironworkers. The guys I worked with were still walking with a lot of pride. I mean, who the hell can walk these beams up in the middle of the winter? I can't! Most guys can't! You'd never see most guys up there. It's a special breed of person.

When I worked at the top, I was doing parapet walls on the edge of the building. We were hanging off the side on guy lines and safety belts. It was our own world. You'd go up there, didn't matter what else was going on in your life. I always say it must be like being in a war front. You forget everything but what you have to do in front of you, and your safety and your partner's safety. When you get a pause, you just take in the magnitude of where you are. You can see forever, you know, forty-five stories up. It's something most people don't get to do. So when you get that pause, you don't start thinking about the problems in your life. You're just enjoying where you are. You become your own country, I think up there. I just started walking taller after being around these ironworkers. I felt I got respect from them. They weren't afraid to hand it out if you deserved it.

— *Kathy* —

Being able to be tough and do the physical work, I got a lot of satisfaction out of that. Going out in the summer and just sweating your ass off and working as hard as you could. In the wintertime here it gets cold. I could always deal with that and not be one of the people who called it quits and went home for the day.

— *Nancy* —

I knew whether I'd done a good day's work or not—there's the conduit.

— *Kathy* —

My mother used to worry about me a lot, walking on bridges across rivers, beams, and all that kind of stuff. And she used to get so upset with me because for Christmas and birthday, she'd want to buy me some nice pretty purse or sweater, and I always wanted a new power tool. She'd say, "Give me a wish list or tell me something you want that doesn't have to do with construction." I'd say, "But Mom, that's what I want. I don't want a new blender, I want a new cordless drill."

— *MaryAnn* —

From an early age I recognized that tools had a magical appeal. That they were very special and that only certain people got to use them. I always was very, very aware that my brothers got the tools, whether it was a chemistry set or whether it was getting to work with my father to put in a new set of stairs. I deeply resented—and I'll try to get over this sometime in my life—at Christmas one year when my brother, who was a few years older than I, got a chemistry set and I got a toy broom.

My brother Jim would take the time to show me how to do some real basic kind of things. I remember we had this big garage. We collected a lot of driftwood from the beach and we'd cut it up to use in the wood stove. I was hacking away at the stuff and he showed me how to use a hand saw so it doesn't screech and it doesn't make any noise. To use it like a musical instrument.

— *Diana* —

I'm working down at Bartle right now. I took Pat down to see it one day and he was impressed. A lot of big machines. And for Patrick, it's worked out well with his school.

They have a Discovery Day once a year where parents give demonstrations or talks on whatever their occupation may be. I did a plumbing demonstration. I had a little fountain that I had built out of copper fittings rigged up with a little valve on it. For every group that would come in, I would just add on pieces so they could see how to solder. I had a little round tray with a drain hole. I set the little fountain in the tray. I had a little hose and hooked it up to the faucet, so they could see the water coming out and going down the drain—the whole principle of gravity drains the water away.

I would let them come up and touch the tools—pipe wrenches and a striker to light a torch with. And then I had the little Presto-lite bottle to do the soldering. I had made a big poster of a house, like a cross-section of a house, with the bathrooms and the kitchen, and showing how the lines in the house run, and eventually they run down out into the street, into the sewer, and how that all works. And that was fun for them to look at. Patrick was in fifth grade. I was a little nervous at first. I thought, They're going to think this stuff is boring. But, I think because I was doing something, and I had the fire and the solder—that got their attention right away and when that happened, I knew I was on the right track. And then they got to stick their hands in the water.

— Kathy —

My kids know which bridges in town are mine. I have a lot of pride in that.

— Deb —

I went to the Mystic River Bridge. They were sandblasting it. I thought for sure they were going to try to scare the daylights out of me. One kid says, "Deb, we can't let you climb over a four-foot ladder. You have to be 18." I couldn't climb any of the steel or do anything with them. I had to be a ground crew.

The day I turned 18, up I went. They put me to the top of the tier in a sky climber. I thought, for sure, they thought I would get scared and just disappear and to hell with this stuff. I loved it! I was right up there, it was great. I mean, the wind would come and slam that staging. I got scared, but even my partner did. I liked the height. It didn't bother me. I just liked being up there.

— Karen —

I never thought that I would like heavy highway. I always thought that just looked like too much work. Those are massive structures and kind of scary-looking. It was not unusual to be walking on a four-inch I-beam a hundred and twenty feet over nothing. It's either railroad tracks or a river or just a highway. And no safety net and no safety lines to tie off to.

I get real lazy if I'm working on the second floor next to the edge, because I know that, Aah, it's possible I could die, but more than likely, I'm going to break something.

But put me twenty stories up and on the edge . . . I'm still going to work just the same, but you have that little excitement going there that—if you fall, you're history, you're not going to come through that one. It's a challenge.

— Maura —

Almost any woman I've ever known who's done some welding has been surprised at how much they liked it. There's something about it, something about the rhythm of it. To do it well you need to really have all of your senses working. The sound, the smell, the look, seeing the color changes and the liquid puddle changes—you have to observe all of that to really do a good job.

— Gay —

Taking a thin piece of metal covered with flux and fusing together two other pieces of material. And knowing that it's perfectly safe. And that you've actually created a third material. And seeing that weld go up there—there's a great deal of satisfaction to that, because it is not something that everyone can do. People can think that they are welders, but there are welders and there are welders.

On the skinner boilers, the men were having trouble not blowing holes in it. Most of the guys, when they weld, want to push it right in, and skin casing is this very fine steel, very, very thin and really almost delicate. It's exactly what it says, it's a skin casing for the insulation and everything that's around boilers. It takes very low heat and it takes just skimming over it. If you push on it with any kind of heat, you blow holes in it.

I had a better touch. I was able to weld lighter.

— Gloria —

They showed me how to cut glass pipe. You have to cut it from the inside. It's almost like cutting wallboard, how you score it and you break it. With glass pipe you cut it from the inside, and you take a torch and you heat it just a little bit and it cuts right where you scored it from the heat. It was different from what I had been doing up to that time—cast iron, galvanized stuff, that kind of thing.

— Bernadette —

I love the smell of wood.

— Cynthia —

The nature of the materials that we work with, the fact that we can bend it, shape it, mold it. The malleable nature of the copper.

— Paulette —

What fascinated me most was the beauty of it before it gets covered up, or on the commercial jobs, where the pipe stays exposed.

The pipe is cast iron black and the bands that hold it together, the couplings, are silver steel and the fittings are shaped at perfect angles. Ninety degrees or forty-fives or sixties or Y or whatever-you-have. To just stand back and look at something that's functional look so beautiful—and to be able to say you did it!

Copper water lines, big 2- and 3-inch ones, smooth, without all the junk around them, smooth and wiped off and clean—it looks like artwork, it just looks like artwork. And it's functional, I like that part. Plumb. Plumb makes a difference. If it goes in crooked, somebody's gonna spot it.

Crawling under houses I don't really like. But I know when I get done they have a new water system that's strapped tight. It's not gonna move. It's not gonna leak. It's fine.

— Gay —

I especially liked—say there was a problem, right, and you had to actually look at something and figure out how you were going to do it. More than, put A to B, but to take time to actually do a small amount of fabrication on the job. Being able to identify a problem and how I would repair it or how I would make it work. To figure out what materials you needed, how you were going to do whatever it was, and then achieve the final product.

— Cheryl —

You could work in the state of Ohio, you could work in the state of California, you could leave the country and go work in another country. You can go anyplace and use this knowledge, and it's going to keep you employable. Regardless to whether you work in a union, or whether you freelance. When you're out there on your own, you can still generate the funds to support your family. I have two children and I'm the sole provider for them.

— Barbara H. —

When I go into these houses—like now I'm doing apartments getting transferred from incandescent lighting to fluorescent lighting—some of the older women,

Working in an electrical closet, electrician Cynthia Long tightens a coupling on a pipe elbow.

maybe say 40 or 50 and older, they're so excited and pleased that a woman is in there doing it. They feel relieved, at ease, because I guess they feel threatened by the guys. Some of them have said they have never seen a woman do any of this type of work. And they're proud.

— Randy —

On 2 Union Square I worked layout for all the glass that had to be set on the whole building, which came in panels spanned floor to floor. By the end, when they finally took away the manlift, the pieces had to all fit. The last thirteen pieces had to fit in there, fold in. I like that part of the trade, that it's got to be right and it's got to look right. I've worked on a lot of high-rises in Seattle. I can drive by and say, I put the face on that building. It's something there to look at, to see.

— Helen —

Starting on something from scratch, from the ground floor up and topping it off—I don't know of any purer delight than that.

— Irene —

I mean, you get to carry tools. You get to hit people, you know, if you have to. You have so much more personal freedom than you would in so many other jobs. Good pay, good hours, pretty much equal rights as opposed to a lot of white-collar work.

I like being able to act out. I like having screaming fights from one ladder to another, and in some ways being the only woman in an all-male crew can be great (I never was offered another woman, so I have to say this, you know). Being a woman among many men, it's like you sort of stand for all women, and so you have a chance to project a lot of thoughts out there—not that you change anybody.

I enjoy the debates and I like to think that I'm adding. I always say to the guys, You are so lucky to have me on the job, you know. You all agree with each other all the time, it's so boring.

— Gloria —

It was an exhilarating feeling once I got used to climbing and going high and plugging into the camaraderie that men experience with each other. It opened me up as a female, that all-male environment, made me take myself less seriously. I learned to joke around and tease and have a good time. I know I'll never be as good at it as some guys.

I mean, there was this one guy, he was such a character, he used to be late to work, he was a drinker. They asked him one day—they were going to fire this guy—Tommy, why are you late again? He says, "I got up this morning and I was on my way out the door but my wife was in such a mood I had to make love to her before I came to work." Everybody just fell out. He could just come out of the blue with stuff like that. He never lost his job. It's a freedom that women don't experience in their jobs. You know, you go to the office, that's it, it's not the same kind of a feeling.

I am just not the kind of person that can be confined in an office eight hours a day. It's just not me. I'd go crazy. I have to get outside. I have to feel the wind in my face, you know? I like the fresh air. Even in the winter.

— Angela —

I like the fact that men can be very direct about what they're thinking. Sometimes I feel like women just want to process forever, you know. About so-and-so's feelings. And how *this* might have happened. It's like everything has to be dissected down to the minutest little thing. And men are totally the opposite, or a lot of men.

Sometimes that can be really refreshing.

You can just say, Look, I don't agree with you, Fuck off. And they're not going to go off and be hurt forever and not talk to you or something. Sometimes I feel like, what I like about working with men is that they are a lot less complicated.

And the physical thing, men are more comfortable being physical. They play more physical sports. If they're working together, they'll rub shoulders, push each other around and I like being able to do that.

— Randy —

Men when they work together are very competitive I've noticed with each other, but in a good way. Competitive to get the job done. But women working together, I think they feel competition and it's not in a good way.

Men, they'll be racing to do rods up and they'll be laughing with each other as they're racing. Well, two women wouldn't be laughing with each other. "God, she's getting ahead of me."

— Angela —

Even though I was raised real kind of Victorian female, I was also raised to say what I thought. I wasn't raised to just shut up and be a complacent woman. And the whole thing of apprentices—female or male—the journeyman is always right? Well, I never agreed with that. If I really felt like, Well, I have a good idea too, I would share my opinion. A lot of guys appreciated that, I think, because—and that's another thing I like about working with guys—a lot of guys are trained to work together. There's a team-type thing that you do together, if you can find a guy who can get beyond the bullshit. And sometimes women have a hard time with that. They think you're trying to take away their power, or they're just not used to working as a team so much, they've been more isolated.

— Melinda —

The four women that were the first class to graduate, we went and we rented a ballroom in the Grammercy Park Hotel. We had tickets printed up. It was $30 a seat. Sit-down dinner and open bar and dancing for four hours. All women apprentices got tickets at half price, and their dates had to pay full price.

We decided that we deserved it. We worked hard for it. It was something like, hey, nobody else is going to pat you on the back, you pat yourself on the back. And you have your family there and your friends there to acknowledge this.

There was a magnificent turnout. It was 200 people. I'd say, at that point, there was about 20 women apprentices in the local and then the four of us graduated. We did a big circle with a picture of all of us. And this one woman, her lover designed these corsages with a big light bulb in the middle. I had this big beautiful red light bulb with all these flowers around it.

— *Irene* —

My favorite event of the whole era was the Blue Collar Fashion of Fantasies Show. We rented the Machinists Union hall and we had a fashion show that was all our fantasies on stage. One of the women carpenters built these huge backdrops, profiles of women, these giant tradeswomen silhouettes. People designed their own outfits and had their own music in the background, and we would come out on stage.

I was the emcee. I had a hardhat with grapes cascading off of it and I was wearing like, long underwear with a push-up bra. I guess I was the Dionysian carpenter. There was an electrician who was like an S&M electrician with BX hanging off of her, kind of like whips, and there were two women electricians who had wired their tool belts so they flashed on and off lights. You know, they were like disco. And my friend Kathy, she had a hardhat that was like a breast. It was formed in the shape— very realistic, you know—because that's how she felt on the job. That she was just this exposed breast walking around. And there was a sort of psychedelic painter and there was a bricklayer who came in being rolled in on a scaffold, in yellows.

And we had a dance afterwards and we had lip synching. We had like three or four hundred people come. We put up flyers in the street, all over 14th Street, Union Square. We wanted everybody to come. Anyone who wanted—and people made their own decisions about their co-workers—but we advertised in the *Village Voice*.

It was such a great spirit, taking on the culture and transforming it into something that we could really deal with and have fun with.

CHAPTER SEVEN

Exceptional Men

I would say the majority of minority men that work in construction here in this area are laborers. And some of those old laborers were the best buds and the best teachers I had out on the jobsite. And they were the best supporters when I was having a bad time or obviously being harassed by somebody. Not that they would go to bat for me or anything. But when they had the opportunity, they let me know that they understood and they'd give me pointers on how to survive. Older black men, cool dudes.

—*Kathy Walsh, Kansas City*

I've always been fortunate in that every job that I've been on there's always been one guy or two guys that have kind of taken me under their wing. That's pretty much how I made it through that apprenticeship.

—*Gloria Flowers, Cleveland*

A lot of male bonding is female trashing.

—*Maura Russell, Boston*

In my first year as an apprentice, Bill Swanson was not only assistant director of the apprenticeship program, responsible for placing us on jobs, he was our classroom instructor for Code. Known for a volatile temper, he seemed to live on coffee and cigarettes. Slamming a book onto a desk or jumping up on a table to scream at someone were part of his routine. In the first weeks of school Bill found a reason to use the word 'tit' in his lesson about every two minutes—"easy as tit," "tit job," "cold as a witch's tit." . . . It felt like an intentional provocation, a reminder that we were six women in a class of ninety, in a local of 2,500—and any adjusting would have to come from our end.

But this man who at first seemed so hostile—making it clear we would be given no easy breaks—had himself come into the union as an outsider, no one's son or nephew. The women in my class worked hard, studied hard, went to union meetings, and took the craft he loved seriously. He was open-minded and bold enough to notice. Over time he became one of our staunchest and most outspoken supporters.

When I sat in his office crying at the end of my third year—one year left in my apprenticeship—there were, as usual, "No if's, and's, or but's about it," as far as Bill was concerned. I'd developed a chronic repetitive stress injury that had temporarily crippled both wrists and left several fingers unable to open and close; it had kept me out of work for six weeks. When I recovered and returned to my job, I was laid off. I was in Bill's office looking for a next assignment, but unsure whether or not my body could handle it. My confidence was low. Bill just stared into me and said, "You're getting your license and you're graduating from this program. What you do afterwards is up to you, but you're finishing this." He sent me out to one of my best jobs as an apprentice, a major renovation at an auto plant where I could thread and bend conduit and learn motor control wiring. But there was no coddling for my recovery. It was an overtime job that went 58 hours a week.

A year later, when five of us became the first women to graduate as journeylevel electricians from our local, Bill was by then Director of Apprenticeship. He had his picture taken with us at graduation, and always kept it on his office wall.

—Susan

The first women who survived and succeeded in the construction industry did so with the assistance of men who had the courage to break ranks with a history of exclusion. These exceptional men, whom tradeswomen speak of with deep affection, often came in unlikely packages. For Nancy Mason's first job as an apprentice electrician,

I WAS ASSIGNED TO BURKE ELECTRIC and the journeyman's name was Patrick Costello. I went into the building and tried to find him. He actually met me and I said, "Have you seen Patrick Costello?" He said, "No," but it was indeed him. He had me running around for a while. The two secretaries in an office across the hallway finally told me that it was indeed this guy I had met when I first got there in the morning.

He had me bend pipe the very first day, and I knew nothing about EMT or a bender. What's interesting is that I did exactly what he showed me to do. We were stubbing in boxes in steel stud walls, so I was bending all the pipe connecting all these boxes. He came by and was watching me and he said, "You didn't ream the pipe."

I said, "I don't know what you mean."

He said, "You have to ream the pipe."

I said, "Well, you didn't show me that." And he got all bright red in the face, started yelling at me.

And then he stopped and he goes, "You don't understand what reaming the pipe means?"

I said, "No, I've never even seen a piece of EMT until today, and you just showed me how to bend it. I cut them the way you showed me to cut them and I've been putting them in."

He said, "You have to ream the pipe or else you'll tear the wire."

I said, "Well, do I have to take them all apart?"

He said, "Well, yeah."

And so I said, "It wasn't my mistake."

He goes, "I know that. But you're going to have to take it all apart anyway." I remember that very clearly. He's a currently retired member, and every July 2, just like it was July of 1979, he sends me an anniversary card as my first-day first-ever journeyman wireman. We had our moments, but he was actually a really neat guy and he really did teach me a lot. He—even from early on—told me that I was smart and that I could figure things out. I got a lot of positive reinforcement about how I would probably be able to do okay. I didn't experience a lot of harassment that some of the other women who I was in contact with were experiencing at the same time. I experienced that later, and I think that I survived a lot of my first year because I was with him a great deal.

On Kathy Walsh's second day as an apprentice carpenter, one crewmate stood up to the ironworker who pushed her down the muddy embankment. This same carpenter reached out to Kathy with his family:

WE ENDED UP HAVING A very long-term friendship. I've lost track of him now, but I got to know him and his family and he got to know me and my kids. I can remember going over to his house and he had a wonderful workshop set up in his basement and him teaching my kids and me how to use rasps and files and stuff like that. We made a whole basket of Easter eggs out of wood with his lathe and power tools. He was great. If it hadn't been for that one person that I could mentally hold onto during that job, I don't know if I would have continued at that point. He was a decent person. So one out of a crew of ten.

For women who faced opposition on entering the industry and came in feeling on the defensive, first encounters with a journeyman, foreman, or steward either confirmed or contradicted their worst fears. Cynthia Long's first job lasted barely two weeks, yet

THAT FIRST JOB WAS VERY important to me in the sense that it did establish for me a base level of my expectation. I didn't know what other people were like, so I just had to base it on this interaction with the general foreman from the first job. One of the reasons that I think he had some sensitivity is 'cause his eldest child was a daughter—I think he may have had two daughters.

He specifically assigned me to a journeyman that he believed would teach me and that would not give me a hard time. If I had run across some of the other foremen that I have since run across in the industry, they would have assigned me to some guy who's like the worst sexual harasser around, put me in the 8-foot ditch with the 5-inch galvanized pipe, and really just busted my chops. He was trying to make it possible, so he assigned me to work with *this* guy. And this guy said, "Okay, this is a blueprint. See this arrow? This is north. See that building, the Empire State Building? That's north. The East River is to your right, that's the East River," and he pointed out certain landmarks. He said, "Any time you go into the building, figure out which way is north," so that gave me a big clue. That was the best advice.

And then he showed me a wall cable. "We're going to pull wire." And he showed me the prints and he said, "See this? And it's got these slash marks? This is how many wires we're going to pull." He showed me. And that's what we did for the morning, we pulled wire. And then he says, "Why don't you splice out the wires. But don't put the Scotchloks on, leave them open so I can take a look at them, and then we'll go from there."

He left me in this room and I spliced them out like he said. Since I had had the background with air conditioning and refrigeration I knew how to make a splice. He came back and he says, "Oh, you've done this before?" He was a nice guy. It was essentially a hassle-free environment for me. And I think it lasted all of maybe a week, two weeks, before they sent me to another job. And then I worked with this really cantankerous curmudgeon.

Even such a brief positive experience could be pivotal to a tradeswoman wanting reinforcement for the soundness of her career decision. Luck played a significant role. A good first partner or crew could act as a shield against hostile situations that followed. Cheryl Camp returned several times as a journeywoman to work for the shop where she began her electrical apprenticeship.

JUST TO START OFF WITH them, it was a great introduction into the trade. They taught me a lot, 'cause I didn't know how to use basic hand tools. And my first journeyman was the shop steward. He had been in the trade at that time, like twenty-three years, so he had a lot of experience and knew how to teach.

What was funny was, they called him "Silvertongue" because of his profanity. But because we were working together and the way I carried myself, he didn't swear around me. No cussing, he was a real gentleman. It was a great job! That helped me develop a mental attitude.

Similarly, the treatment tradeswomen received from union officials was often critical to their decision whether to become involved in union politics and activities. It also served as an indication of the long-term employability they might or might not expect in the industry. Diane Maurer was an apprentice electrician in IBEW Local 46 in Seattle, where both membership and leadership by women has been among the most successful of any construction union local in the country.

I GOT IN IN '79. I think it was the next year, in 1980, there was a change and the Jordan administration came in. He had some staff members that were real supportive of women being involved, and I think that's made the difference. He had one business rep, Hank McGuire, he was my journeyman when I was a first-year apprentice, and Hank is a progressive person. He's the kind of guy, whenever he talked about his wife, he always referred to her as Lee Anne, and he always said her name like with such respect. He loved this woman, it was obvious just from the way he said her name and the kinds of things he would talk about. Whereas you work with other guys and it's "the wife," "the old lady," or "my wife"—something that you *own* or something that's *there*—there wasn't the same kind of appreciation. Hank values everybody's intellect and it doesn't matter if they're a woman or black or whatever. I think he was pretty key to encouraging women's involvement.

I worked with him in establishing our political action committee. He was very good about taking a large project and breaking it down to specific tasks and saying, "Oh, if you could go take care of this, it would be really helpful." Off you'd go, and he'd give you the people to talk to and the kinds of questions to ask and you'd get some information, and then you'd be off over here finding out this and that and the other thing. He was always quick to say, "Good job," and quick to give you a recognition at the union meetings so people knew who was doing what. Then the women that got involved, they didn't have horror stories, you know? They could say, "Yeah, it's worth it."

Tradeswomen like Karen Pollak and Melinda Hernandez, whose fathers or grandfathers trained them to use tools, could carry a confidence in their mechanical skills learned in childhood onto the jobsite with them. Melinda, who grew up in the Bronx, recalls her father's open-mindedness:

MY FATHER, WHEN HE FIRST got his first car—he got a used car—and when he used to work on it, I used to watch him. I was about 14. He never told me to go inside, or I'm not supposed to be here, I'm a girl. My father, he's very mellow. He's got his own ways about him that are macho, but he never rejected me as far as, if he was doing something and I wanted to help him or I wanted to learn. He would never push me away. My father has a fourth grade education and he comes from the hills. So he was learning himself, but while he was learning he was teaching. And he would say, "Pass me a screwdriver, pass me the pliers," you know. "No, not that one, I need the Phillips," and things like that. I kind of just worked with him. I'm talking about in four-degree weather, when it was freezing out and we had to change a fan belt because the fan belt broke. I'd stand there with him in the cold and we'd do it together.

Even now he calls me up if he has problems with the car, and I go over and I help him. So that was the bond him and I established at a young age.

The foremen and journeymen that tradeswomen speak of most fondly are never the ones who gave them easy tasks or told them they could just sweep the floor, the pay's the same. They are the skilled craftsmen who held them to high standards and who—in exchange for hard work—were willing to share the tricks of the trade they'd accumulated over a lifetime. These exceptional men were critical to tradeswomen's developing the competence and confidence to become journeylevel mechanics in a competitive field, with the potential to run work. Lorraine Bertosa recognized her good fortune in becoming a core crew member for such a foreman:

I WORKED WITH ONE GUY OFF and on for the last few years that understood that camaraderie real well. One of the best carpenters that I've ever seen. He was a carpenter foreman and I was building houses with him. Bob would always try and gather the same bunch together. On the last job I was on with him we had forty-five carpenters. He had his little core, but he had to hire all these other guys.

He would set the mood and we'd work off of him. People had such respect for this guy. Lunchtime he'd have the popsicle truck come in. He'd set a real playful tone. When we worked, we worked. He expected us to work. He expected us to be there early. He expected us to stay late if we needed to, not to race out the door at four o'clock. Because if he could break you free, he'd let you go. He'd always give you breaks if he could—as long as you were willing to work with him. He expected a lot of you, but he played.

He created such a phenomenal environment, you'd give your eyeteeth to work for this man. To be on a job with him. Just your eyeteeth, it was such a treat. Made you feel like a professional. And that's what's gone out of the trade. Hardly anyone cares now. That's the real shame in the whole thing.

A lot of respect for his workers. In three years I only saw him yell twice. The two times he yelled had to do with a safety situation. Never yelled. One time a buddy of mine and I, we were going to put up these pink Formica bookcases. They had put fourteen of them in the room, eight feet high, four feet wide, real narrow. They just put them all in the middle of the room. We had to go around, screw them into the wall. I remember Len picking up one. He just touched one next to it and he almost decked me, because three of them, they went down like cards, one went into the other one. They busted up.

Bob came in and he surveyed the situation and he says, "I think you can fix this." Just like that. When his boss came in and says, "What do you need this pink Formica for?"—'cause Bob had to order a sheet of pink Formica—he said, "Oh, we had a little problem." Infamous in the trade is, let's blame somebody. This man was so above that. He showed me that you can have this wonderful work environment. You don't have to work like animals. You can get more work done than anything.

And he pushed the snot out of this guy who ran the outside wall crew. Bob started giving me these porches to put on these houses, and I didn't know porch for shit. So I'm struggling to get through these porches, and this other guy is going, "God, it's taking her a long time to do these porches." So Bob put *him* on a couple of porches. And it was taking him just as long. Bob says, "Just shut up, it took you that many days to put that porch up." He'd do that. He'd nail 'em down.

He's the only one that ever asked me to be a foreman. He wasn't that way just with me. There was a black man on our crew and he said, "Bob's the only one that ever made me a foreman either, Lorraine." And Bill was a phenomenal carpenter, real skilled at running a crew. But he was black. Bob was the only one that dared to try us out.

I look at people like Bob, or I see men that aren't phased by *who's* in there, not phased by color, as long as it's a good worker—the guy knows he's a good carpenter. The guy knows he has work whether he gets laid off or not. Your health comes from people who feel confident that no matter what the hell they're gonna do they're gonna be all right.

Ironworker Randy Loomans also found that the men most able to value her were exceptional in their own skill and self-confidence.

I've HAD A LOT OF good mentors. One of the best layout men in the Northwest here, I've worked with or around him for the last five years. Now I've become his partner basically. Any job he goes to. And we work doing layout with a transit and level. It requires thinking more than anything.

This man who's really kind of like a genius, I've worked with him and he makes mistakes. We laugh about them, I know even the best make mistakes. We always have the joke, "Measure the cloth seven times, cut it once." There's nothing that's so monumental that can't be fixed, I've learned. If you cut something off too short, you just get another piece and weld it back and—my goodness, you've got a full piece again.

Still, mentoring relationships, particularly between a man and woman, can be emotionally complicated:

I WAS CUTTING THESE SHEETS, I cut a little far. And I said, "Well, faceplate'll cover it." He just went off on me. I just looked at him and I said—I know him so well that I can say anything to him—I said, "You know, Herb, if a man would have done this same exact mistake you wouldn't have said shit. You wouldn't have said nothing to him." He says, "Well, I expect more from you." In some ways that's good and in some ways that's bad.

The next day he's back onto this whole subject of what I cut. Before work I said, "Herb, you're big-dealing this to death. I can go out and fix what I just did." And he goes, "Okay, enough said." He knows when he's gone too far with me.

I think he expects more from me because he wants me to be so much more. It's a good expectation. I never think of it as negative. But the idea that I get reamed more than somebody else would get reamed, I don't really like that. But this is a man that I've mentored under for five years. We're in a different plane than most people.

Making mistakes in construction, though sometimes costly, is an inevitable part of learning and working in a trade. Yet tradeswomen's errors were often responded to, not as normal and individual, but as proof of the gender and racial stereotypes that had supported decades of exclusion. Being respected and accepted enough to goof up sometimes made an enormous difference in lessening the stress load. Paulette Jourdan, used to being yelled at for mistakes, appreciated the foreman she had right after she'd graduated from her plumbing apprenticeship.

OH GOD, I FLOODED A couple of rooms. He blew up right away 'cause, I mean, somebody was gonna have to pay for the damage. The computers almost got damaged, but luckily the water missed them by a couple of feet. He's throwing his hardhat down and cursing. I was piping. It was just this small cap on a three-quarter inch, I forgot to solder it. I mean, I had run all this 3-inch pipe and nothing leaked, and they turned the water on and things were fine, so they closed the ceiling up. They'd just sheetrocked the ceiling.

Once he found out it was a cap I had forgotten to solder up in the ceiling space—that guy came to me later that day and apologized for yelling. I was crying, and he said, "You think I haven't made mistakes?" He said, "Everybody makes them. Everybody makes big ones. And you know, you can't kill yourself over it, just forget it." He was very sincere, it just blew me away. That would be what I would expect a woman to say, to have that kind of attitude about it. He was just a gentle soul, just a real sweetheart. Rare, but they are out there. That's how I got through it. Guys like that.

In a harsh industry where maneuvering through dangers is a necessity of the job, behavior that is in fact kind and respectful might seem otherwise to an outsider. As an apprentice carpenter, Karen Pollak could appreciate the actions of her foreman when she froze up one day, working on a bridge.

WHAT WAS INTERESTING ON THE heavy highway job was, there are days—and it didn't matter who you were—there are days where you would just get freaked out about being out on those beams. You could walk them the day before with no problem. But you got up that morning and you took two steps and just went, I can't do this. It's a real strange thing that happens. The eyes are really, really big and glassy. You're not paying any attention to what's going on. Whatever you grab hold of, you've got a death grip on it.

It happened to me only once, luckily, but I was out in the middle. That morning I'd been walking the beams just fine. No problem. That afternoon, after lunch, I got out there and I went to turn around to do a layout and I just couldn't move. Everybody was yelling and screaming at me, "Don't look down. Just look at me and come back here."

It was like, I'm not moving. There was one laborer that was real good friends with me and so he came out to get me. My foreman told him to leave me alone, that I had to get back on my own. Most of the time, if you froze, they would take your hand and kind of walk the person back sideways.

It was like, Okay. Fine, I'll get back there.

I got down on my hands and knees. This was a four-inch beam. It was a 1920s bridge, the one that goes between Missouri and Kansas and crosses the Kansas River. At one end of the bridge you're probably sixty feet off the ground. At the high point you're probably a hundred fifty feet off the water. In fact, when that bridge had been built originally, one of the superintendents had been killed because he had fell off the bridge in the exact same spot. We had just had a truck driver get killed the day before.

You're standing sideways. You just kind of inch yourself around and go, Okay, and start trying to squat down. If you go one way or the other, if your nail apron shifts or if your hammer drops down behind your knee or something, it's going to knock you off balance. So that's scary.

Those people were good people. He was doing it deliberately, but it wasn't malicious. A lot of times, you try at first to get the person to come back on their own.

There are times when you know that it's not going to work. That they're just going to stay there and if you don't go get them, they'll be there for a week. After it had happened, I pulled in guys lots of times.

As in any work situation, sexual attractions developed on construction sites. Tradeswomen would get many offers for dates, sex, and Chinese food. But getting involved with a co-worker or foreman made an already vulnerable apprentice more vulnerable. Painter Deb Williams broke her own rules against dating someone from work, with a boilermaker on a power plant job.

H E ASKED ME TO LUNCH. I said, "I bring my lunch." He said, "I do, too. We'll meet and we'll go out in the parking lot." So we used to meet and we'd walk out in the parking lot. They're all, Hey, Bill, hey, Bill. They used to tease Bill all the time because he used to charm me with honeydew melon, he used to bring me honeydew melon every single day. I'd finish my lunch and I'd eat the honeydew walking back to the jobsite. "Oh, Deb got a honeydew, Bill must have brought honeydew."

He's what they call a pusher, a subforeman, and where I was working he could come out of his area, come find me, talk with me for a while. He asked me out on a date. I told him I was a widow with six kids and I didn't have the time and all kinds of excuses. And he was, "Who cares?"

What was nice about Bill is that they never knew I was dating him. I didn't date him for a very, very long time while the job was going on. It was a couple of weeks before it was over, I started dating him. I was interested. So I agreed one day—I must have known him five, six months or so—to meet him to go for pizza. Ever since, I've been with him eleven years.

My girlfriend Julie had dated some guys on the job and then break up with them. And they've always got something smart to say. Billy's very quiet. I don't know why he's with me, but—one gabber, one quiet. He never told any of those guys on the job anything about me. They asked him if he was dating. He used to do this thing—closed fists. You know, None of your business. And he would never talk to them. So I never got a reputation from them. But it was really weird because they never knew I was dating him till one Saturday they had picked him up here for something and I was coming out of the house with stuff. They were like, Wow, I didn't know you was dating that girl painter. That was like a year later. He had enough respect for me.

Even a simple, small gesture of kindness can take on enormous significance in an environment where it's unexpected. Helen Vozenilek found that expectations of foremen or journeymen became "a self-fulfilling prophecy."

I REMEMBER ONE OF MY LAST jobs as an apprentice. I got out to it and the guy that was the foreman, he welcomed me, and he said, "I know you can do this work because I've seen you do it before." To have someone say that to you—it's like telling a first- or second-grader that they're going to do good. It's like a key to success, I think.

It was ironic. I had worked for this company twice before and had always had a terrible experience with them. I felt like they had treated me pretty rotten, so I had a lot of trepidation about going out to this job with this contractor. But for this guy to greet me in that way, it sort of set the stage for me to do well. And I did. I ended up running up the job at the end just because this guy had expressed that confidence in me. There's no doubt in my mind that there's an exact correlation between encouragement and success.

CHAPTER EIGHT

Balancing Alone across an I-Beam

from PAST THE FINISH LINE
for Sara, Jill, Cathy, and Margaret

We survived isolation by the law of mutual induction:
      magnetic fields of bodies
      separated physically
      can still overlap and empower.

Where a 100-year curse had vowed
'no woman shall pass here'
we passed:

all five
      as one.

                                —Susan Eisenberg

— Sara —

I was watched. That was the thing. Every second of every day within the jobsite there was one pair of eyes on me. At all times. Just out of total either amazement or wishing you weren't there or I don't even know. Not always a hostility—a lot, just an interest. But what I felt and how I took it was pressure.

We had to be really, really extra special good, the women. And really push ourselves hard time-wise. I'm kind of a go-getter worker anyway, I wasn't really much of a fart-around-er. But I also didn't kiss ass. I had already done factory work and I knew how they could own you. There was no way I was falling for that act, but I also did want to make a good impression—it was a fine line to walk.

Graduating apprentices Margaret Gove, Sara Driscoll, Cathy (Cunningham) DeRosier, and Susan Eisenberg at the Local 103 IBEW union hall on May 5, 1982. Photo by Francis J. Daly, Jr.

I remember every Friday I would come home and sob. I mean, really let my guard down. Because I had just spent the week being on display. That first year was exhausting—not only physically, but emotionally.

— *Irene* —

I was the first woman in my local. In fact, there was a "house of ill repute" across the street from my local called the Carleton Arms. And when I got initiated—we're initiated and then we went down to the meeting—one of the guys from my local who I got to be friends with later on told me he thought I was a girl from the Carleton Arms who had just somehow wandered in. So that's how unusual I was. Meanwhile, I'm wearing like, cowboy boots and jeans. I wasn't exactly dressed like, you know, a Lady of the Night—well, maybe I was, I don't know. I thought maybe I was dressed kind of macho. But anyway, that's how unusual it was in my local.

— *Helen* —

I got some support from friends, but they didn't really understand what was going on. Unless you're on the job you don't really get how it feels, you don't really get all the nuances. I would go home and just feel like I was lower than whale shit.

People would see me in my outside life and say how competent I was. And I would say, I'm not competent, I can't do this, I can't do that. It was hard for people to see that, because they couldn't see me in that environment that was, a lot of times, overwhelmingly negative and unsupportive. People thought, Well, the answer is obvious. Why not get out?—which is a friend kind of advice.

But I didn't want to quit. I guess at a certain point, that was my test, to make it through. Because, if I quit, how am I going to feel about myself? I'm going to feel like I couldn't do it, I wasn't good enough. I didn't want to leave not feeling good enough.

— Barbara H. —

My mother, my sisters, my aunts, they all have been strong backboned women. My aunt, she was the first black woman sergeant-lieutenant-deputy-chief-of-police-and-all-that-good-stuff in the D.C. area. I figured, if she could hang where she was, I could do what I had to do here.

I moved back home for the first or the second year of my apprenticeship. I had moved out when I was eighteen. My mother didn't think I was ready to move out at eighteen anyway, but I was determined to be one less burden on her, because there was seven of us. She had always held down two to three jobs until she turned 50. She was a charwoman and now she's a charwoman inspector, you know, cleaning up for the government, now she inspects for them.

I had to be to work at, what, 6:30, 7 o'clock? She would get up in the morning and make sure I had a hot breakfast before I went to work. That was good. She would tell me, "This is what you really want to do—you go in and you do it. You have to believe in yourself. I believe in you. Hang in there." She encouraged me. Anything I wanted to do, she told me I could do. If it got hard, she would let me cry, and she said, "You finished now? Get on back up in there."

Sometimes I get to talking and tears come to my eyes, but it doesn't mean that I'm falling apart. I try not to let that be seen at work, try not to show, I guess, my feminine side. I wanted to prove to them, No matter what you said to me, I was still coming back the next day. I didn't get into a lot of verbal arguments or heated disagreements. Because I believe my work will prove who I am and what I am.

— Gay —

My support system's been me, basically, only because there weren't support systems there. It's real goddamn hard as a tradeswoman to talk to someone who is sitting at a desk, who may have been in her job for ten years, who has *no* understanding of what it is to be the only woman on the job with eighty-six men and know that most of those eighty-six men don't want you to be there. It's freezing-ass cold and you've done your best to do a good job and you get the layoff check at the end of the day. And the guy who's been standing over there doing nothing but warming himself is on that job tomorrow.

It's very hard to explain to someone, I did my very best but I still lost my job. Most people, when you say you've been laid off, "*Pfffft,* you mustn't have been doing your job." We work with that all the time, that somehow or other we are last hired, first let go. It's always a challenge to know how well you are actually doing. There is nobody out there that can say to me, "Gee, Gay, you did a really good job."

So what do you do? You internalize it all till you get to the point where you say to yourself, Am I doing an adequate job? Should I be doing this? Is it possible that I have misjudged myself and I'm not up to doing this? You begin to have all kinds of self-doubts.

— MaryAnn —

The woman who ran the coffee truck, she was funny. The first day I was on the job, she pulled up—and she was glad to see a woman. She told me that the supervisor had asked her if they could use *her* name and her daughter's name and put them down as employees of the construction company. She said, "Sure, if you give me— how much do you earn a week? Give me a full paycheck and sure you can use my name. Otherwise, forget it. Hire a woman." They were deeply offended that she didn't want to help them.

She was the only woman that I saw all day long. I saw her at coffee break and I saw her at lunchtime.

— Paulette —

When I first started, I'd go sit and eat lunch by myself and read a book. Well, one guy looked over at me and he said, "What's your problem? You don't want to eat with us? We ain't good enough?" I thought, Oh, my God. I figured, I don't want to hear this shit. Let me just go sit with them, even if I'm not listening to them. 'Cause I didn't want to make waves, I'd just started. I didn't know I had rights and also, even though you have rights, you *are* in that environment and you make it as comfortable for yourself as you can.

I'd see women in other trades occasionally. But one time when I had lunch with one, talk started going around about us being together. Even though I'm not gay, I didn't mind *that* so much as that—you can't do *anything.* You can't even have lunch together and people are starting shit. We weren't even sure if we were allowed to leave the jobsite, because you don't want to sit there in their faces and have lunch. But we still had lunch together.

— *Kathy* —

I think it was about my second year at school, they had a Symons forms demonstration. Symons is a brand name for concrete forms that come in panels, it's a whole system. Everyone had to come to school and get certified and it was an all-day Saturday thing.

I'll never forget walking in there and seeing other women. All of us were like, Oh God, there's another woman here! I had met one or maybe two others at that time and knew that they were out there. But I never saw them or was able to hook up with them. We all went to this all-day seminar and it was like, Oh, they fucked up bad. Because we exchanged names and addresses and phone numbers. And that was when we first started getting together and meeting on a regular basis. There were five or six of us.

— *Donna* —

We had a women's caucus in the carpenters union. There was a group of probably a dozen women carpenters that met regularly. We would have potlucks or get-togethers where maybe up to forty women were involved. It was really because of that group that I got through my apprenticeship.

The other women carpenters became my best friends. I knew that whatever I was going through, somebody else was going through across town. We helped each other find work when we were out of work. We set up a network to share information about where the public works jobs were, that were required to make some attempt at affirmative action hiring. And we were active in our union locals.

— *Sara* —

Luckily we had our little women's group in the car. Four of that first group of six drove from Jamaica Plain out to the school together, two nights a week. That car pool was the thing that allowed me to stay with the program, because that's where we would get to whine, cry, scream, holler, moan, bitch, and laugh and tell tales on what was going on on the job. The fact that we car-pooled like that really saved my life. I think it saved all of our lives if I recall. Various ones of us had our shirt collar pulled up saying, Oh, no, you're not leaving now, we're all in this together. Come on, hang on.

— *Irene* —

There were a fair amount of women being taken in as apprentices during that period of time. There were women around the school and that was where we sort of got to feel our oats a little bit. It was almost like a girl gang, you know, there was support there.

— *Cheryl* —

I was the very first woman to go through Local 38's apprenticeship program. A trail-blazer. They said that there was one female before me that started, and I think she worked for maybe two or three months and dropped out.

RTP contacted me to congratulate me for getting into the apprenticeship program and to let me know, if I had any problems, that I could come down and talk to them. Recruiting Training Program? That's an outside entity of the union, that helps to bring minorities into the different local unions throughout the city. They had women's groups, a rap session. I went to a couple and I didn't go back. Maybe for those that needed to vent their frustrations it was good for them. But I really needed someone to tell me, "I've been through the four-year program, and you're going to run into this and run into that—but it's okay. If you just do this and do that, or if you have this kind of attitude, then you'll make it."

But all of us were starting out, brand new. There was a couple female carpenters, a female sheetmetal worker, a couple plumbers, and we were all at ground level, just feeling our way blind. Didn't know what we were going to do or how to get around some of the obstacles. All we could do was talk about what happened during the week and how we managed to get over. And maybe this would help somebody else if they have the same problem. But there was no experience there to learn from or to draw from. We were all the first ones, so we had no one.

It was basically just a gripe session group, and you felt more depressed when you left than you did when you came, so I stopped going.

— *Irene* —

A couple of us presented the director of apprenticeship with the idea of having special training about sexual harassment, for the school itself to have it brought up as an issue and discussed. We even came up with a lot of ideas of how it could be done—but basically he was discouraging about dealing with the issue of sexual harassment in any organized way. I think we were intimidated by his attitude and just sort of gave up on that approach. From that time, we kind of retreated into our own networks.

— *Randy* —

I worked on the Cedar Falls Dam, and there were several women up there. That was the best group of women I've ever seen. Working women. I mean there were truck drivers. There were pile bucks. There were carpenters, laborers. It was a busting-ass job.

Us women hardly ever got a chance to talk to each other, that was the sad part of it. 'Cause women don't do that on jobs. It's just back to this old saying, Three men could sit and talk for an hour, but you don't dare put three women together standing huddling and talking. Because you're so out there. There are so few of you that you're spotted immediately. We don't take that luxury with each other. Men do it all day long, every day.

— Marge —

Two women get together on a job and then things are driven between them all the time. Suddenly there's this external gossip. It happened with me and a steamfitter on the hospital job and it caused hard feelings between us. And I kind of fell into it, too.

I had been in seven, eight months, I was the old lady on the job. Oh, God, she came to work first day—lot of makeup, wrong kind of shoes. They'd comment. And I wouldn't say, She's got a right to dress however she wants. I'd probably be more likely to say, Yeah, I don't know, that's weird. It's not like I went, Nyah nyah, she looks wrong, or anything like that, but I did not do everything I could have done for her.

— Helen —

Just because you're women, doesn't mean you have that much in common. You might have stuff in common about how it is being on the job, but you have very different ways of dealing with it. I never felt a lot of solidarity with other women on the job.

For one thing, you always had to be aware because the men would look askance when the women would hang out together. There's a dynamic set up in the trades, too, I think, with women, that somehow you're in competition with each other. That somehow we can't all be good, that someone has to be exceptionally good and the other ones are, you know, not very good. Either subtly or not so subtly, that dynamic plays out when women are on a construction site together. I would feel that from other women, that they wanted to shine by themselves.

But, for the most part, there weren't that many women.

— Randy —

Most jobs I'm the only woman. That's what amazes me, these high-rises that get built. I've been on jobs where I *am* the only woman. On a fifty-story building. I mean, something's wrong.

— Sara —

I would get isolated by some of the straight women who would come on the job from other trades, who would spot me as a lesbian. As much as the boys didn't quite know what to do, the women generally could figure it out pretty fast. And they would back off and not be friendly, and that was hard.

— Deb —

I used to get the wrong impression of some of the women coming in. It's like, If you can't work as hard as me, what the hell are you even doing here?

I worked on a power plant with a lady carpenter and a laborer and an engineer. They all had easy jobs. I'm working like a bull next to these guys, and it used to tick me off. And the guys would say things to me like, Did you know she was with this one and she was with that one? All the backhoe guys used to talk about the engineer. One said to me, "I dated her and I got all I wanted."

After a while I got sick of listening to it and I said something to her. She said, "Deb, he's bullshit 'cause I wouldn't go out with him. That's all a lie." She wound up—with me standing there—confronting him, and he goes, "Well, you know how guys talk."

— Paulette —

The first lesbian there—they gave her a really hard time. She was leaving as a I was coming in. She was the first woman. She was very open, very out of the closet. Stories would go around. She actually did publicize the fact that she was getting married. I'm thinking, Why are you giving them all this ammunition? But 'cause I couldn't know what she was going through, I never could say to her, I think you ought to stop. Besides, the damage was done before I got there.

She did graduate, and she did work a couple years after. But they did give her hell. She was ready to take them on—I really admire her for that, 'cause she was out there by herself for a year or two. They just really had their way with her, anything they felt like doing. I have no idea what all that was, 'cause she rarely talked about it. I just heard a few things from some of the other women.

They wouldn't train her. They beat her down. She left, just got out altogether. She doesn't do plumbing, just doesn't do it. She went back to nursing. I hate them for that.

— Mary —

At Busch Gardens, I remember this one black girl out there, she was working as an ironworker and she was really new. She was climbing up a ladder and this guy grabbed her ass. She complained to the steward. His remark was, "This is construction. The guys are going to be guys." I was too new myself to come out and be argumentative, I really didn't even know what to say to the girl. We were walking together and the steward was telling us, "There's not much I can do, I'll talk to the guy." And that's it. I'm sure they all had a big laugh in the shack about him grabbing her ass.

She left the job and I assume the trade. I never saw her after that. It's really made me angry to see all these women drop out.

— Deb —

I never went to a barroom with the guys off the job after work. I never went to a barroom with them at lunch. I figures, just totally avoid the whole situation. Not that I was a drinker and I wanted to go for beers, but it seemed that's where the in-crowd was going. I felt like I was missing something.

The guys that stayed behind would go to the parking lot at lunch and sit in their cars and, "Come on, Deb, sit in the car, it's more comfortable." I'd sit on the hood of the trunk or on railroad ties or whatever, so everybody in that parking lot could see me.

If you sit in a car with four guys, old pig minds start rolling. If I was in plain sight all the time, they had nothing to talk about. I used to say, It's summertime, sit outside. Then we'd have a whole group. They'd sit out there with me and all have lunch instead of everybody sitting in individual cars.

It was twelve years later before I ever went in a bar and had a beer with one of the guys at lunch. It was like, if I wanted one I should have had one. But my worst fear was the reputation for something that I didn't even do.

— Mary —

If you went drinking with them, then you'd be called a slut or something. That's out of the question.

— Irene —

I was on one job, 80 Broad Street, right in the middle of the boom where there was about six women on the job from different trades. While the guys would be hanging out watching the secretaries—it was all construction jobs all around—we would have our own little enclave and we would be outrageous.

We went to the tradeswomen's convention in California. We walked off the job, there was three of us from the six. They were doing overtime and we said, "We don't care, we're going to this tradeswomen's convention in California. Should we take our tools with us?"

We had our jobs when we came back. And the guys all were saying, "We didn't see it on TV, it wasn't on the news." They thought we would be on television, you know, because we made such a big deal out of it. They thought, if you're turning down overtime to go to this, it must be something that's going to be on the news at 7. It's got to be something that big.

But—to be able to fake power—I loved being on that job. It was collective power, and also you weren't Everywoman. You got a chance to be an individual. There was some variety where the pressure was off, because you didn't have to be The Woman's Opinion.

— *Helen* —

I ended up going to this conference in Tucson for tradeswomen, I think it was '81, '82. That was my first understanding that, Oh, I'm part of a movement. I read about it in a women's magazine or something, and then three of us from Albuquerque went down, a carpenter and the other electrician that I knew. I just wanted to go meet other tradeswomen and maybe talk about what our jobs were like. For me it was new and it was exciting. It was women that were just first in—I think a lot of it was sharing stories and coping mechanisms and what do you do. Because at that point, a lot of us were apprentices, not that many journeywomen yet, so we were, in a way, at the same place.

— *Karen* —

We had heard all these strange things about people on the East Coast and the West Coast. They were supposed to have lots of women in the trades.

I went down to a friend of mine's bookstore one day and it was just this magazine sitting on the table, *Tradeswomen* magazine. I thought, Well, maybe I'll subscribe. In those days you might get one magazine a year. Maybe two, if you were really lucky. There was a time when I had sent in my money and then months and months later—here'd come the magazine. And you'd go, Oh yeah. I forgot that I paid for that.

— *Irene* —

A group of us who had gone through Women in Apprenticeship and stayed together as friends and met and strategized and everything—a whole bunch of us carpenters—we decided we were going to write the president of the district council a letter and ask to have a meeting with him to discuss the fact that women have entered into his union for the first time, and that we would like to have a session with him where we could talk about the issues that have come up. We didn't know him from a hole in the wall.

We thought we were so smooth, and we thought, well, how do these union guys deal with each other? They always go out for dinner and drinks. So we said, We would like to join you at the top of the World Trade Center as our guest. We thought somehow this was like this really cool thing to do, and we discussed what we were going to wear. Of course, we thought he would accept it and we would get him up to the top of the World Trade Center and we would talk to him and get to know something about him and have his ear for a while, as a group, not as individuals, you know. What happened was, we got a letter back saying, Refer all problems to the head of the Apprenticeship Program. Basically, what are you bothering him for? So we thought, well, I guess we can't play ball. I mean, we thought this was just a terrific idea.

Then shortly after that he was found—*he* wasn't found—the union car and his wallet was found and he was never found. He was shot. I mean, he wasn't interested in incorporating women into the union! They never found his body. Somebody did away with him.

— Melinda —

We came through, we were the first ones, and we took a lot of scars for a lot of the women that followed. Whether they liked it or not we were there and we were visible. And we made an impact. We met like once a week, and we formed this group, this small club called Women Electricians. And basically it was just the four of us getting together and supporting each other and talking about what was going on on the job. We were a strong network at the time because we all were fighting for the same cause and the same battle. And what happened was, as we got more women coming into the union we tried to recruit them into this club. Because we figured, you know, in numbers there's strength. But a lot of the women were forewarned that this was a group of radical feminist women that were creating problems in the union.

To me, because we take a stand and we say we don't like the fact that we don't have a pot to piss in on the job—that's not radical. We take a stand because we don't have a place to change our workclothes—that's not radical. If we take a stand because somebody comes on and sexually harasses us—that's not radical. All these things are just fighting for your rights.

But to them, it's radical, because you're making them face the fact that you're here to stay. And not on their terms, on equal terms.

SETTLING IN

THE "INITIAL" PERCENTAGE GOALS AND timetables for women's work hours on construction jobs and for women's participation in apprenticeship programs established in 1978 under President Carter were never increased. Neither were they reached. During President Reagan's administration, budgets for federal enforcement of affirmative action were reduced.

In 1996, according to U.S. Department of Labor statistics, women were 2.5 percent of the construction workforce.

CHAPTER NINE

Bucket or Bathroom?

I didn't go as often as I might have. Or when I did have my period, I probably changed my tampax less. A lot of times because it was a pain in the butt to get to a bathroom. Sometimes it was too damn cold to even pull your pants down, it was like soooo cold. So, varying reasons why going to the bathroom became an issue. Distance. Quality as a bathroom itself—you'd avoid it at all cost if you could!
—*Sara Driscoll, Boston*

Since the dynamics on that particular job were already fairly hostile and I did not want to attract any extra attention to myself, I decided not to tell anyone at work that I was pregnant. Which on most jobs would have been difficult because, in my first trimester, I needed to pee all the time. But luck had me working on the finish stage of a luxury hotel where I had my choice of twenty-eight clean bathrooms on each floor. And there was always toilet paper.

—Susan

Sanitation facilities on a construction site are the general contractor's responsibility. Bathrooms, like coffee breaks and quitting time, were an issue that came up on the first day—and provided a good indicator for what could be expected about conditions more generally.

What for workers in many occupations is a small personal matter in construction can become a very public issue. The nature of the facilities tradeswomen encountered varied widely and depended on location, what was being constructed, the phase the project was in, the quality of the general contractor, the strength of their union representation—and attitudes toward women.

The question of toilet facilities raised the issue of whether or not the industry would accommodate the needs of female workers that were different from men's. Women have additional reasons for using a bathroom and usually need more time when using it. Even the fact that women sit down to urinate makes using a johnny-on-the-spot more uncomfortable, especially in cold weather, and more complicated, as Barbara Henry found:

I HAD JUST BROUGHT MY TOOLS, I think it was maybe the second or third day. Went into one of them johnnies-on-the-spot and I wasn't thinking. My tools just went *ka-plook*. I lost screwdrivers, sidecutters—Kleins. Ooh, man, couldn't get nobody to turn the thing over so I could get my tools out, because back then I didn't have that kind of money.

When the guy came to suck it all out, he said he didn't see anything, so I figured it had just gotten sucked up in the hose and went on its merry little way. So I learned not to take tools into the john.

After a while, I learned not to go to the john at all, because in the wintertime they don't dump it as well as they do in the summertime. I had conditioned myself. I would not drink anything during the course of the day. I would eat in the morning before I came to work and make sure I went to the restroom then. Lunchtime I would get in my car and go to the restroom. And at the end of the day and that was it. Other than that, I couldn't go into the johns because they would get so funky.

An advantage of working on a remodeling job was that tradeswomen could expect to have access to bathrooms used by the regular female employees. These were likely to be clean and warm, with such luxuries as soap, toilet paper, hand dryers, even a tampax machine. There were other problems, though, for women dressed in construction clothes and work boots, as Boston electrician Sara Driscoll discovered:

THE THING ABOUT WORKING IN an office building and having to use the ladies room—you have people scream because you're walking in there. Or not scream, but look. Or jump. Or "Hey! You can't go in there, young man."

I'm like, Oh, yes, I can. I belong here, I'm a girl, don't worry about it. It happened to me a lot in the Prudential Center in particular. And then, 'cause I was there for two years, finally they got used to it. I could go to the bathroom without having somebody tell me that I couldn't.

When appropriate facilities *were* provided for women on the construction site but not for men, tensions festered. Special bathroom shacks or port-a-johns for women at times became the focal point for anger at women's presence on the job and were vandalized. Or they became the setting for confrontation, as happened to a female traveler on a job with Melinda Hernandez in New York City.

I REMEMBER WE HAD TWO BATHROOMS at the convention center. When I say two bathrooms, it was one bathroom for about 600 or 700 men and there was only about ten or eleven women, so the men kept pouring into the women's room, too. So this woman who was a traveler from out of town went to use the bathroom and this guy came in. She said to him, "This is the woman's bathroom." He said, "Well, I see men coming in here all the time." She says, "At least wait until I finish, and then you can use the bathroom." He says, "No, I can't wait, I've got to get back to work, so I'm going to go now."

So he went into the stall and used the bathroom and when he came out she said to him, "You know, you're no gentleman." And he says, "Fuck you, nigger. And if you don't like it, I'll kick your ass." To make a long story short, we tried to force the steward to get this guy either laid off or—he gave her like a half-assed apology after his collar was pulled. But he should have been laid off or transferred. There should have been some immediate action taken against this guy to show that this behavior is unacceptable.

So what happened was, we got together and we drafted this letter. She was a black woman from Texas. And when I took a stand for this traveler, she got canned, and I got canned, and they canned this other woman, just to make it look like it wasn't a personal thing.

As an apprentice ironworker in Los Angeles Mary Michels often worked on the initial phase of jobs, well before there was plumbing brought up to the floors. Only on the roof, reachable by crane, and on ground level were there regular port-a-johns since (it was claimed) they were too large to be brought up to the other floors on the man-lift.

W E DON'T HAVE NORMAL OUTHOUSES on high-rises. Do you remember the half-houses, the half-bathrooms? It's an out-house cut in half. It's opened up on the top. If you stood up, everybody could see you, and if anybody walked by, they could see in. Well, that was a problem out here.

I got a few bladder infections from holding it so bad, because I couldn't go down to the bathroom—it would take too long, to go down a man-lift when you're on a high-rise. Then I stopped drinking coffee. Then I stopped drinking water. It got worse. So what we did in the beginning, basically, is just push the outhouse where nobody was at. You have a big floor there. For years I had to push it somewhere else.

I got to go fast, so I can get in there, go, and get out so nobody walks up on me. It's embarrassing. I used to run from floor to floor to see where there's nobody at, a lot of running. And then different times, I would have to have someone guard for me, if I knew and I was comfortable with him. "Watch, make sure nobody comes by, I have to take a pee." The guys don't care if you're there or not, they go right in front of you. Sometimes they'll turn it so you can't see it, but I often thought, boy, I could get a lot of these guys on indecent exposure.

After many years in the trades, Mary began to see "thirds" on the jobs, outhouses with only the top third missing, that did give more privacy.

Difficult access to reasonable bathroom facilities was not only demeaning, it was a health hazard that could cause medical problems and add stress to an already stressful job. It also made women vulnerable to criticisms about lost worktime and productivity. Sometimes this was intentional, and became a means of harassment. On her first job as an apprentice in Kansas City, Karen Pollak was initially told—and she believed—that the job had no bathrooms. When she moved up to the next floor, she realized that there were bathrooms on the odd-numbered floors. But, although she was the only woman on the job and had just started, they all had signs on them that said MEN ONLY.

B UT I HAD USED IT ANYWAY. I had just dropped the nail down, so that you couldn't open the door. One of the foremen had wanted to use the restroom and he couldn't because I was in there. He waited for me to come out and told me that it was for men only, if I wanted to use the restroom I had to go to the top floor. That's where *my* bathroom was.

It was probably up to sixteen, seventeen floors by then. I had to climb the ladders, because we didn't have stairs. The elevator couldn't be used except for moving materials and for the bosses. I watched all the other *men* in the trades use the elevator all the time, but no, I used the ladders.

I used the ladders one time and went all the way up to seventeen, eighteen floors to go to the bathroom. Then I got in trouble because I had been gone too long. It was like, "Then you probably should let me use one of the MEN ONLY restrooms." "Nope. You're just going to have to learn to be faster." I have to be back in a certain amount of time or I'll get fired, that's the way they told me it was. They expected me to climb up the ladders and run to the bathroom.

Leaving the job to use a bathroom elsewhere was one solution to inadequate facilities. Though women could usually do this at lunchtime, it meant, essentially, breaking conditions: going to the bathroom on the worker's own time rather than on company time. And sometimes, such as when a woman had her period, waiting until lunch when the job started at 7 a.m. was just too long.

On the job where she lost her tools in the johnny-on-the-spot, Barbara Henry had a foreman who understood the problem and let her, when she needed to, leave the job during work hours.

I F IT WAS THAT TIME he would let me get into my car and—there was churches up the street—he would let me ride up there and take care of myself. That was nice of him because a lot of guys would say, I have a bucket right here, you can do it right here in the bucket. And I'm like, No, I won't be doing it in no bucket. That's why when you're on a site you try to find out where you're going to be and if you can drive, you're going to drive yourself and find the nearest facility around. You know, a carry-out or something.

I believe they should try their best to get running water on all the sites, even if it's just in a assigned spot where you have to lock it up and every female on the job gets a key. They need to have that, because that hygiene—guys don't care, but women, it's very important to them.

If bathroom facilities were not adequately supplied by the contractor, it was the union's responsibility to represent the interests of the workers. Cleveland plumber Gloria Flowers:

I HAD ONE INCIDENT THAT ACTUALLY went to the business agent. I was over on the West Side and they had these two heads out there. They were so filthy I couldn't go in 'em and hold my breath long enough to use the bathroom. I tell you, the honey-dipper couldn't get in there soon enough to service that thing. So I told the foreman, "I'm going to this KFC I've seen down the street around the corner." He told me that I was not to leave the job to go to the bathroom.

I had to go.

He says, "I don't care where you go. You're not leaving this job. You can go in a bucket over there in the corner, but you're not leaving this job." And I said, "Well, you do what you have to do, and I'll do what I have to do." I was an apprentice, in my second or third year. And I left that job and went to the bathroom.

One of the guys called the business agent, because the guys got incensed when they found out how the foreman was treating me. It was like having friends that you don't know you have. It warmed my heart, it made me appreciative. I couldn't remember half those guys if I saw them again, they probably wouldn't remember me. But one guy told the rest of the guys, "Hey, if the girl's got to get in her car and go to the bathroom, she should be able to do that. If she can't go in the head, then she should be able to go. It's only a five-minute drive away, what's the big deal?" I remember him saying that.

So that went to the business agent and he came out there. He talked to the foreman, he told him to get off my back.

In a job setting where the dignity of a worker is respected, basic physical functions like urinating and defecating, having a menstrual period, or being pregnant should not be accompanied by anxiety. As Melinda Hernandez put it:

THIS IS RIDICULOUS, THESE ARE basic things that it's not even too much to ask. I'm not asking you for a vanity in the bathroom, you know. I'm not asking you for a stereo system in the shanty. I'm asking for a place to change and a pot to piss in, to put it vulgarly. A simple thing.

In New York City, contractors traditionally provided workers with a changing room for switching between their street clothes and work clothes. Electrician Cynthia Long noticed the difference when this issue was quickly squared away in a forthright and respectful manner.

THIS LAST JOB THAT I was on, it was the best work experience that I have ever had. What made it the most enjoyable is when I asked about the sanitation facility and the changing facility, the attitude was that they were going to take care of it. So there wasn't all this struggle that I usually have to go through about why I deserve to have some place to change, why I should have a clean sanitation facility to use. That made it possible for me to focus on the work that had to be done.

CHAPTER TEN

Carrying Weight

Painting was a lot easier than working in the pizza shop. I only
worked a couple hours on the register, but I set the whole place up.
I had to make barrels and barrels of pizza sauce. I had to make three
or four 6-pound cans of tuna fish, make salads. It was back-breaking. I used to say, I'm really earning my money here, I really am.

Painting a room, setting up a ladder, and going around brushing
was a lot better than fixing tuna fish.

—Deb Williams, Boston

Angled at a perilously steep pitch, the twenty-foot extension ladder still barely reached the
I-beam. One of us would have to climb it, hold the pipe with one hand, and with the other
shoot a clip into the beam using a ramset gun, a powder-actuated tool with a strong kick-
back. John chose himself and asked if I thought I could foot the ladder in case it slipped or
the ramset blast threw him off balance. Yes, I nodded (though really I wasn't sure).

Both of us were sweating as he got into position to fire off the first clip. Just before the
shot, I heard him mutter, "Nothing's holding me but a wing and a prayer."

—Susan

In construction, strength does plays a role in both productivity and safety, though
how significant a role depends on the jobsite and the assignment. Having the physical capability to handle the job is an issue that tradeswomen have grappled with
and considered since they entered the industry.

Pre-apprenticeship programs designed for women often included a component
on building physical strength and endurance. Many tradeswomen joined a gym,
lifted weights, or worked out to make sure they were in shape to do the job. When
she was starting out, Lorraine Bertosa read about a woman who "used tomato soup
cans to start improving her capability of lifting. I did that and within weeks I found
that I was much more able to do the work."

With her crew moving switchgear weighing several tons: Sara Driscoll (far left).

Although the caricatured image of construction workers idealizes brute muscle, in reality skilled tradeswork involves a wide range of mental and physical skills. The increased use of prefabricated and newer materials, such as plastic pipe replacing cast iron, and the increased use of mechanical lifting devices, have reduced the relative importance of brawn. According to ironworker Gay Wilkinson, whose sons and husband are also in the trade:

I VERY DEFINITELY FEEL THAT WOMEN can be ironworkers, especially with the technology now—you don't have to lift the piece by hand. There is some kind of machinery or equipment to help you. The whole piece about, you have to be as strong as a bull, really doesn't exist anymore.

Rod people, it's harder. I believe that you get to a certain age, man or woman, and you should not be lifting Number 18 bars to do rods, and very few older men do. These are the jobs that they send the younger apprentices to do. They're the ones that are out there lifting the Number 11's and things like that. A Number 11 is usually 1-inch diameter and it's probably 9 to 12 feet long and weighs unbelievable amount of weight.

But when you get to a certain size in rods, if there is equipment available, they will have that equipment pick up the bundle and move them, and then you do it piece by piece. Unless a woman were *really tiny, tiny*, she can hold her end. There are young male apprentices that have trouble doing rods.

I think the barriers are all mental. I do think that if a woman wants to be an ironworker—or actually any of the trades—that we should do a lot more about building our upper body strength. Now, I don't mean running out and doing weight-lifting so that you'll look like Joe Jock, but I think we have to be more aware of our upper physical fitness. I'll put most young women I know that have a baby that weighs about twenty-five pounds, with her diaper bag, up against any young man I know, because, without realizing, by lifting that weight all the time, your arms do become very strong. But if you don't have that situation, then it's just a matter of developing it.

Mary Michels, a second-generation ironworker, expanded those ideas:

IT'S REALLY MADE ME ANGRY to see all these women drop out, and I know a lot of times they'll say, Well, the women can't cut it. They can't do this, they can't do that. But, you know, on the same hand, everything out in construction is based on male, and I don't understand this.

If something's too heavy for a male to pick up—get a crane. If a woman can't pick it up—she's too weak. That is a whole attitude out there. I don't think it's right. If it's too heavy, make it smaller, you know. It just amazes me. They don't see it. They don't see it.

Although apprenticeship programs had no entrance requirements testing strength, nothing prevented many journeymen and foremen from inventing their own "tests" to weed out women they deemed unqualified. Bernadette Gross took many such "tests" during her carpentry apprenticeship in Seattle.

I WAS WORKING ON A ROOF. They usually cut the tiles to a point, about two feet by two feet, and you throw them. Well, as soon as I got on the job the guy says, "You can stay until 12 o'clock. If you can't hack it, you're gone by 12:00." And for that first four hours, they cut those things about twenty inches wider and just really made it very difficult for me—they're more like four feet, you know.

Right after lunch I said, "God, these things feel smaller, lighter." They say, "Well, this is the size that they really are." Because I didn't even know what was going on.

Over time she could place this treatment in the context of the workplace culture, concluding that "about fifty percent of it was gender-based. Any new person, any low man on the totem pole, was going to get the shit. Depending on your reaction to it—whether they had found a button, they would continue to push until you got so frustrated that you left."

Having worked with her grandfather, a master cabinetmaker, Karen Pollak had the confidence in her skills and clarity about her limits that comes with experience. Other female apprentices in Kansas City lacked that advantage, and were more vulnerable to taunts about their strength.

A LOT OF THEM GOT HURT. That was, I think, the saddest part of the whole thing, because they were trying to be macho instead of smart about it. They would just go fall right into the trap. It would be like, I'm picking up two forms—you ought to be able to pick up two forms. If you can't, then your ass is down the road.

Instead of saying, "I'm sorry. I'm not as big as you are. If you think that that little guy over there can carry any more than I can, well, then, bring him over and we'll match." But I can't carry anything compared to someone that's six six, that weights 280 pounds, that lifts weights. It's like, Right. I'm going to pick this up and do what? This one woman she ripped her shoulder muscle, wound up getting a shoulder replacement.

One of those women in Karen's class who found it hard to set limits was Kathy Walsh. A large-framed woman, six feet tall, the apprenticeship was her first exposure to the trades. After her first job lasted only two weeks, she explained to the training coordinator that, being responsible for three children, she was concerned about having steady employment. Four months later he sent her to a second job assignment, expected to last a year and a half, a water treatment plant being constructed from the ground up.

THIS COMPANY, THEY BROUGHT ME into the office, filled out the papers, showed me a movie on safety, told me what their policies were on harassment and discrimination and OSHA. This is the color hardhat you wear and this is who you'll be working with and this is what you'll be doing—big union contractor, big job. Took me out to the crew I was going to be on.

After they had left, the foreman, he's like right in my face saying, "I never worked with no fucking woman and I ain't never going to. I will run you off of this job before this week is over."

I didn't say anything. I just made it up in my mind that I didn't care what they did or said, he would not run me off before the week was over. I was determined to do the best job that I could. And if I quit it would be because *I decided to,* not because of him.

We were building these huge radius gang forms, must have been like 40 by 30 feet with radius red iron. We were nailing on the plywood facing and building the forms that were going to be used on this huge round water treatment plant. He'd have the biggest and strongest carpenter on his crew and me go carry these I-beams. And it was really weird because the week before I started this job—I had forgotten about this—I was having some uterine problems. I had had a laparoscopy where they went in through the belly button and looked around. They found out it was nothing serious, but I still had the stitches.

I was out there and he had me doing the heaviest and nastiest stuff, all the grunt work. The second day out there I ruptured my stitches.

Kathy didn't stop work to take care of her injury, but bandaged and taped it when she got home.

I WAS AFRAID THAT IF I gave them a reason that they'd get rid of me. It didn't heal very good but it didn't get real bad. I have more scars on my hands from that job than *that*. They had me nailing Number 10 coated sinkers. Coated sinkers are a bitch to nail if you're not a good hammerer, 'cause they bend and the heads on them are real thin and flat. So I had blisters on top of blisters on top of blisters on my hands! It took all I had just to survive work every day and go home and survive kids every night, plus going to school. I can remember coming home from work and being so tired, and Zane was little and sitting in his room in the rocking chair and it was like—the only thing that felt good was sitting in that rocking chair with him, rocking him to sleep at night.

Historically, unions have spoken out for the model of skilled tradesworkers who practice their craft as a lifetime career, contrasting this model to shops that employ one skilled worker to direct the labor of many "young bucks" hired for their brawn and replaced once they're injured. Construction unions have recognized that even a large and strong person has physical limitations and have fought, both in contract language and "traditional practice," not only for health and safety standards to minimize injuries, but also for the economic protection of older and disabled workers.

The entrance of women into the all-male terrain of construction, however, created a situation where union workers and leaders had to either reaffirm their commitment to the standards of a safe workplace and the trades as a lifetime career or—actively or passively—"break conditions" in order to pressure women into leaving. On Paulette Jourdan's second day as an apprentice plumber at Livermore Labs, she confronted a pipe that weighed slightly less than herself.

I GET ASSIGNED TO THIS GUY. We jumped in this beat-up old truck, drive 200 yards to where the pipe is stacked. He jumps out and goes to pick up one end, and I jump out and go to the other end of the pipe. I'm starting to pick it up so we can take it over to the truck and he says, "No, you pick up your own." He was such a bastard! Bald-headed, nasty, mean guy.

Four-inch cast iron pipe weighs about 90 pounds and I was weighing 105, and I had never picked up, first of all, anything that heavy. And it's awkward and it's stiff and it's in ten-foot lengths. And this was in the mud! I couldn't believe it.

I didn't know that I had any rights. I didn't know what was fair. I thought, If I don't do what they say, I'm going to get kicked out of the program.

Four-inch pipe, you don't have to pick it up by yourself. I didn't find that out for maybe six months or a year. You're not required to do anything horrendous. I mean, if you're stupid enough to pick up 200-pound pipe—*if* you can put it on your shoulder, *if* you're a man, *if* you want to break your back,— if you're stupid enough to do it, sure, they'll let you do it. They don't care. But they don't *make* you. They've sent guys on jobs by themselves to put in 6-inch cast iron, and I know that stuff weighs at least 160 or 180 pounds.

Oh God, it was a nightmare. I didn't know if I could pick up that 90-pound pipe. I was trembling, my whole body was trembling, and my feet were stuck in three inches of mud, but that sucker was up there and I didn't hurt my back—cause I wanted to prove to this guy, Maybe I can't do this job, but I'm gonna try this. And if I can do it, I'm gonna do it.

Nevertheless, a complaint against her was filed with the apprenticeship program by the contractor.

THEY WROTE A LETTER AND said I wasn't strong enough to do it. When they wrote something in my file about my not being able to do something, I always—a couple of months down the road—was able to do it. They'd call you in before the [Apprenticeship] Committee. The first time I got written up was 'cause they said I couldn't handle 4-inch cast iron.

By the time I got in front of the Committee, it was four months later. I'd been carrying cast iron down the goddamned stairs into the basement. And I talked like that, too. "Your letter is old, I already got this shit down."

If upper body strength was in fact a common problem for women that required remediation, nationally coordinated training programs could have adjusted their curricula, as they have flexibly accommodated other changing demands of the industry or student body. Although Paulette stayed in and graduated, she learned that another female apprentice was dropped from the program on similar charges. When women working union jobs were not clearly told about standards for lifting, or when their physical competence was unfairly challenged, a key distinction between the union and non-union sectors was sacrificed.

Raised on a 600-acre farm where hard work was a way of life, Randy Loomans never doubted her physical ability to make it as an ironworker. She was sent to work on the West Seattle Bridge, where three female apprentices before her had failed to last for even half a day.

T HEY LITERALLY TRIED TO WORK me to death. In my whole apprentice-
ship that was the only time I had any trouble where my apprenticeship coordinator intervened on my behalf. It wasn't because I even went and complained to him. I was talking to somebody in the bathroom, and his secretary heard it and then it all got out.

We were packing rebar across the deck, a double mat. Here *we* were taking seven—me and a man, a big man. *Two men* were taking five bars. But he'd go run to the bundle first and get his seven bars up. The rest of my seven would be all entangled, so I'd have to come from the middle and work my way out to get my load. I was doing twice the work he was, just shaking it out. Then we'd go. I'd fall in holes and he'd just keep going. I learned later, just drop the load on him a couple times. There's a bunch of things to learn about it that I didn't know at the time. At the end of this day I was so wore out I couldn't even hardly stand up. I was just beat to the ground.

I was bending over these dowels, a kind of a bar that you put over the rebar and you bend it over—and it was good size rebar. I was moving from one to the other. I wasn't speeding. I had no speed left. The boss came up to me in a real threatening tone and he says, "You better get up on it or I'm going to get your money."

By then I was so disgusted and mad I said, "Well, why don't you." I came down there later and I said, "Well, do you have my money?"

He goes, "No. You going to be here tomorrow?"

I says, "Well, I thought you were getting my money."

He says, "You going to be here tomorrow?" It was like a game he was playing with me. He knew I couldn't quit. See, when you're an apprentice you can't quit a job.

Well, Ray came out on the job then and said, "Just what's going on here?" This was my apprenticeship coordinator. They finally let me alone and let me work and quit messing with me and my mind. I was hired out for one week and I lasted the week. They wanted to keep me, but I had so much rage inside me I would have *never* wanted to work for this company.

I went right to one of the owners of the rebar company and I said, "What is wrong with you people? You know you have to have us women. Instead of beating us down, train us. Train us right. Don't run us into the ground where we'll never be back again. Let's have a partnership here."

But without strong oversight on the jobs and exit interviews with women who left the industry—neither of which were part of affirmative action policies—that contractor was free to assert that he made a "good faith effort" toward compliance. He had in fact hired four women! No record would contradict or put into context his claim that three were physically unable to do the work and the fourth refused a job offer.

Smaller women faced extra hurdles, since the usual way to get a job done didn't always work for them, and materials were sized for someone a foot taller, with a longer arm span and more weight. Electrician Cheryl Camp needed ingenuity when she found herself unable to bend larger pipe.

SOME OF THE GUYS SAY, You know you can't do this. But there's a lot of small guys that are in the local union that probably have a hard time with it, too. It's just in your determination, 'cause I know I've surprised a lot of people with what I am able to do.

I was really small when I first started, you know, 5′6″, and I weighed about 119. I *did* have a problem bending one-inch and inch-and-a-quarter pipe. I can do it now, because I have my own method. I think as women or just as small-framed people, you have to develop your own method for doing things that is right for you, right for your body frame. It might take you a little bit longer than that big burly guy over there, but you could still get it done.

That inch-and-a-quarter bender with the shoe on there? The way I have to use it to bend pipe really looks awkward to someone, I'm sure. You have to jump on the shoe for a little bit to get the pipe started, and then I come around behind it and push the handle while kind of standing on the pipe. It looks like I could fall over any minute, and actually I have to have something to balance me so I don't fall over. But I can bend it without kinking it—as long as it's that softer pipe? You know there's the two kinds? One-inch pipe, when it's the softer kind, I could bend it without any kinks. That real shiny hard pipe?— kink it almost every time. I just don't have the body weight to keep down on the floor to keep the bend going up smoothly on the radius.

You still have some people that they just cannot help that urge to come watch you and then tell you, "Well, no, you shouldn't do it like that. Do it like this." Even though your method will work and it's a fine method. They want you to do it their way.

There's always more than one method. *This is my particular method that I chose to use.* You may do it another way, but it's going to work out the same. I'm going to achieve the same end as you will and will probably have it done just as fast. Or I may be a couple steps slower or a couple steps faster, what difference does it make? As long as I get it done.

Ideally, every mechanic is versatile.

In reality, every worker brings particular assets and liabilities; the strongest crew is one whose workers have a range of complementary skills. Being small and flexi-

ble, having the patience for precision work, having an aptitude for technical information are all valuable skills in modern-day construction.

I'VE BEEN IN SITUATIONS NUMEROUS times where I had to go up above a
ceiling, or down in a manhole, or anyplace where the big guy is not going
to fit. "Cheryl, we have a special job for you."

I've walked across ceilings on the black iron that's suspended. 'Cause they
figure, Well, how much do you weigh? We need you to go out there. It's like
walking a tightrope, you know. I'm scooting along on this black iron, holding
the wires as I go so I don't slip off, making my way across this finished ceiling
so I could get to an open box or whatever has to be done. On the outside looking up at it, is the finished, complete ceiling, and here I am above it.

The real issue, particularly on a union-protected jobsite, should be whether or
not a worker can carry his or her weight in the team's production output. Physical
strength is only one facet. Still, when women came up against physical limitations it
was troubling to their self-esteem and self-confidence. Diana Suckiel became a
plumber when she realized her pay as a licensed practical nurse barely covered
childcare expenses.

I REALLY DON'T LIKE RUNNING THE big pipe. It's not like everybody's out
there trying to be a superman, you know, but there are just some guys who
can lift more than others. On some jobs, if they're doing real heavy stuff that I
can't do and they don't have anywhere to put me, then the foreman has to do
a lot of shuffling around.

Even though we have a lot of things that help us lift the heavier pipe—
hijackers, forklifts, roustabouts (they're like a small movable crane)—there's
just a certain amount of weight I can lift. I tried to work out and get my muscles built up. I felt like I needed to do my job and do what I was getting paid
to do without having to ask for too much help and be a burden instead of
being part of the crew. Just by watching what everybody else was doing and
how much other guys could lift and do, and when they would ask for help and
when they wouldn't ask for help. I knew what my limitations were and the
guys, the longer they work with you, they'll know what your limitations are.

Co-workers' attitudes about women working in the trades naturally played themselves out whenever women needed physical assistance. In addition, since construction is an industry where the workforce is always temporary, where through
different phases of a building's construction the crew size expands and contracts,
workers can feel particularly protective of their specialties. And for some men, their

special asset is exceptional physical strength. Plumber Angela Summer developed an approach to those partners that was usually successful.

I HAVE A PRETTY GOOD FRAME and I was stronger than a lot of guys expected me to be. I would carry around a lot of stuff that was heavy, but— I mean when you get to 6-inch cast iron pipe, some guys will do it, but a lot of guys do expect to do it with another guy.

I did not ruin my back and I know a number of women who have. Sometimes it's hard for a woman, if you're working with one of those guys that *does* do it by themselves to say, "Okay, that's great you're going to carry it over by yourself. But you're going to have to help me with this one." Sometimes I would have to kind of struggle with that and just say, "Look, you know, I don't want to stand here doing nothing, but I can't carry this by myself. We could probably get this done faster with me helping."

I remember one time getting into a fight with this guy because there was this huge, immense pipe that had to be moved up this stairway, and I suggested we move it together. "There's no way we could do it together," he said. I had moved a lot of furniture and I have a pretty good sense of space, I've driven heavy equipment and I do really well with that stuff—I thought we could make it up the stairs together. He got really angry with me, I mean, just really angry. He felt like it could only take one person and he would be the one to do it. He was a huge guy.

Sometimes you *would* get the satisfaction of the other person—maybe not saying it—but somehow you would get the message that they were actually kind of happier that they weren't having to be Mr. Macho! I mean, we were going to get just as much work done and more probably—and they weren't going to have to kill themselves in the process.

Ironworker Randy Loomans put the issue of strength into perspective. Overvaluing biceps underestimates the value of the brain muscle.

I T'S NOT THAT THE TRADE'S so hard—no matter what the image. It's hard. But it's hard for men, too. I don't think every woman can be an ironworker, but that's true of everything in life. I've tried to keep myself in shape strength-wise because it is important. But I think being smarter is more important, being smarter than the piece you're working with. Or using something for leverage. Where men, they've always got to be macho and break their backs.

Ironically, the fact that affirmative action standards were enforced primarily on large federal construction and Department of Transportation jobs meant that women, especially women of color, were often stuck doing heavier, more repetitive and physically demanding work. Many tradeswomen recall apprenticeships where, rather than being rotated through the full range of on-the-job training, apprentice carpenters spent years building and stripping forms, apprentice plumbers spent years coring holes, apprentice ironworkers spent years in the rod patch, installing and tying down steel reinforcing bars.

CHAPTER ELEVEN

Customized Treatment: Women of Color

It's hard to separate.

Because as an Asian woman when somebody reacts badly, I can't tell whether they're reacting badly because I'm Asian, because I'm a woman, or they just don't like my personality. It could be any of a number of factors. I think, though, that on some levels it might have been easier for me as a white woman. And I think that as an Asian man I would have gotten further.

I think that it's a lot of missed opportunities. I certainly was there and available.

—*Cynthia Long, New York*

Though the hotel job had lasted several years and there were not many women on the site, I was unaware of the harassment faced by the black woman laborer there until ten years later, when I saw a black male electrician who'd worked there, too, at a union meeting. He and I had been friends and allies. He asked if I'd heard the good news about the two guys from that job finally getting their paybacks for what they'd done to the woman laborer.

I knew the guys he was talking about, but I had no idea what he meant. I had heard that one of the two had been convicted of murder, but I knew nothing about the other until my friend told me that the guy had been busted for drugs. And the good news was, when he got to prison, "they were waiting for him, raped him, he hasn't been the same since."

I was taken totally aback. Here was a man whom I'd always known on the job to be jok-ing and accommodating—who'd always looked out for me when we worked together— happy from a depth I'd never seen on his face. Because finally, it seemed, there was some balancing out. Of what, I didn't know.

I felt like I'd just been shown a print negative. He began to develop the picture for me.

—*Susan*

With so few women at all in the industry and women of color especially underrepresented, (in 1988 according to the Department of Labor, there were no black or Hispanic women reported in most construction trades) it is difficult to make conclusive comparisons between the experiences of tradeswomen of color and white tradeswomen. This is particularly true since, by nature of the work itself, no situations were exactly identical. Yet the observation was widespread among tradeswomen of different races, regions, and crafts that a woman's work environment, training, and employment opportunities varied with her race and ethnicity.

One of the first four women to graduate as a journeylevel electrician in New York City, Melinda Hernandez, who is Puerto Rican, remembered:

ON THE JOB, WHEN BETH and Cynthia and Jackie and I were coming through, Cynthia and Jackie and I went through a lot more harassment, because Jackie's black, Cynthia's Asian, and I'm Puerto Rican. A lot more was dished out at us than was towards Beth, who was Jewish.

I think that it was harder for nonwhite women to come through. Because I remember when I was a sub-foreman. This number two guy that didn't like me—there was a white woman apprentice on the job, and he would tell her, "You should study and you should become a foreman or a super, try to get into the shop, maybe into engineering." He was telling her that she should upgrade herself. But he wouldn't help me upgrade myself—and that was an apprentice. So, yeah, I see it.

The harassment was in major amounts from all males. But not as much towards the white women. They [Hispanic and black men] wouldn't harass the white women, because they were afraid the white men might come out for her.

Bernadette Gross, when she began her carpentry apprenticeship in Seattle, was the only African American woman in her local.

THAT MADE ME A DOUBLE minority, and that's how they treated me. You're only here because you're a double minority. Nobody ever said that to me, but what they expected of me, or what they were willing to invest in me, was nil. If you're just a part of a crew that gets laid off all the time then— hey, you're just with that layoff crew, it's nothing personal. I know some women who have gotten with the company and stayed and, in all honesty, they were white. I don't know how much that had to do with it.

In San Francisco, carpenter Donna Levitt, who is white, had similar observations:

I THINK THAT IT'S HARDEST FOR the African American in the local to stay employed, from what I've observed, mostly as a union rep. I think that it's uncommon for women to go from one job to the next for a company, although it happens. It's particularly uncommon for the African American women.

Ironworker Mary Michels:

I SEE A LOT OF PREJUDICE towards blacks. They're not rude to the black men in front of their face, but they are when they're gone. You know, the usual remarks, calling them n-i-g-g-e-r-s and stuff like that. A black female—God help her. The black girls—they don't get work, they get ran off, they don't get hired. They don't fire them or anything like that, of course. Work slows up—We don't have the parts here, we got to lay you off. And the next day, they can turn around and rehire somebody else.

As a Cherokee Indian in Kansas City, Karen Pollak faced racism not only on the job but at the apprenticeship school.

T HE VERY FIRST PROJECT I had to do in school was to learn how to hang a door and fix a door jamb. Because the apprenticeship coordinator had said something—"another dead Indian." There was an empty beer can laying in the hallway from last night's school. I was at the water fountain and he was walking past. There was the beer can setting next to the water fountain. As he said that to me, he turns and goes into his office and he slams the door rather loudly.

Okay, I'm pissed. At that particular time I was pretty hot-headed about, if you slammed my race I would really get upset. I went to turn the knob, but I never got the knob turned in time. It was a hollow core door and so my foot went through it, and then the jamb was a real shitty jamb, and so the whole thing just kind of fell right in. I kicked his door off the hinges. Well, that was my first project. We had to fix the door.

Racism created many divisions among tradeswomen. Karen was aware of her advantage over women of color who were not as light-skinned.

I 'VE TAKEN JOBS WHERE, YEAH, I've met a double quota, so they're really happy, because then they don't have to hire any [more] minorities, period. Because I filled this quota *and* this quota. It works out really nice because I look white. And so it's like, the boys aren't going to get real upset. Because they'd a whole lot rather see me than have to work with a black person or—God forbid—some other lower minority group.

And painter Yvonne Valles, a Chicana, observed in Los Angeles:

THE ONLY WOMEN I'D SEE in the union were white women. I'd rarely see a Mexican or a black. I only worked with one other Latina, actually.

If you're blonde, blue-eyed, halfway decent-looking, the guys would just go crazy. Even the Mexican guys they go crazy when they see the white girls. In Spanish we call them *hüeras. Hüera* means light hair, light eyes, light skin. But these guys see the *hüeras,* they treat them almost like they're secretaries sometimes.

When you're off work the union will say, "Go down and see this contractor, get there about 6 in the morning before work starts, he needs a couple of people." This is what I used to feel—when you go into a shop and you ask for work they check you out. They look at you from head to toe and if you're homely, they don't want you. But if you're halfway decent-looking or they think they might have a chance to go out with you, they'll hire you. I noticed that they go by looks a lot.

We had this one black woman I met, one time she was off work and I think they were going to send her down to a job and they sent me instead. I think she got kind of upset because she probably felt, why did I get asked and not her? I think because she was black and she wasn't a good-looking woman, she was heavy-set.

With painting, the rough painters do all the exterior work. Sometimes they won't put them inside offices, 'cause a lot of them can't handle a brush. All they can do is spraying and sloppy work because they don't use drops when they work outside. A lot of times they just slap that paint on. But when you do interior work, especially in high-rises—you're working for these law firms, doing all these law offices and medical offices—the work has to be perfectionist work. I used to notice that the white girls would get a lot of the real nice finish work or paperhanging work.

I think they used to stick the minorities out doing the rough work, meaning it's exterior, there's going to be an outhouse there. I'm not going to be working inside a building where I can use a nice restroom. I did mostly exterior work during my apprenticeship. I rarely got a chance to do interior.

Carpenter Irene Soloway, who is white, noticed similar job segregation in New York:

I THINK FOR BLACK WOMEN, IT'S obviously much more difficult. Black women have been mostly put out on heavy construction, mostly because it's federal. They fill two slots, you know. And I think there's a lot more sexual harassment of black women, particularly by white men, because they feel that that's appropriate somehow. I was on the convention center and there were black women who were trainees. There was definitely a lot more liberties taken with black women. That's what I've seen.

There are some black women that I've met who have been very fine crafts-people and have a lot of integrity and have gotten into the more skilled areas. But generally, the assumption is they get dumped in heavy construction and basically used up.

In Kansas City, when two apprentice plumbers—one white, one African Ameri-can—became pregnant with their second child during the third year of their appren-ticeships, the difference in their treatment was striking. Diana Suckiel, sponsored into the local by a cousin, was aware that her family connections and race had eased her path:

W HEN I FIRST GOT INTO the trade, a lot of guys didn't know who my relatives were, but it did help a little bit. And I think it helped being a woman not-of-color. Because as time would pass, I would see a little bit of a comradeship because I wasn't black or I wasn't Mexican or I wasn't Indian— because I was a white woman. They felt more comfortable around me as I got to be in the trade longer.

Diana appreciated the support and security provided by union membership during her pregnancy. She was on a job where she had to work off a 20-foot ladder; her doctor recommended she stop working after her third month. As guaranteed by the 1978 Pregnancy Disability Act, she was able to collect weekly disability pay through the union from the time she stopped work until six weeks after she deliv-ered her son.

T HEY DIDN'T HAVE ANYWHERE TO put me to keep me on the ground, and I think that my contractor that indentured me, they felt like I might be a liability. I don't think they wanted that on their conscience or wanted it on the insurance. So, because the doctor told me not to work, they said, Okay, you can take a leave, but you have to stay in school.

They just treated me like a human being and a pregnant woman—they were respectful to me, as far as the men running the school and the teachers and everything.

But when another third-year apprentice, who was African American, became pregnant, also with her second child, she was asked to leave school—which meant the end of her apprenticeship.

BEING DIVORCED WITH ONE CHILD and then in her third year, she gets pregnant by somebody—not her ex-husband, by a boyfriend—that just really struck everybody bad. If she had been married and was pregnant, which I was, I don't think people would have felt so odd about it. I mean, they didn't about me.

They didn't say anything to her face, but there was a lot of bad things said about her. Like, What does she think she's doing? She's got this opportunity and she's jacking it up. Making comments about, Well, she's divorced—she's not married. And What is she getting pregnant for when she's already got one child to take care of, and here she is in the middle of an apprenticeship program. The guys couldn't understand it, you know.

Eventually the teachers even talked to her and they said, You're going to have to quit, because we can't have this. To a certain extent, the school has a lot of power about conduct of apprentices, and they have thrown guys out for their behavior. The fact that she wasn't married and she was pregnant with somebody's kid—where with me, being married, I was just continuing my family.

I felt it was a little unfair for them to make her quit because of the way it looked. It was the appearance thing. I really don't know everything that came down because she really didn't want to talk about it. She was very upset about it.

Particularly at a time when hardly any women were above apprentice level, tradeswomen lacked the access to information and power necessary to assist each other through difficulties in the process, and the personal effort and focus demanded for any woman to succeed left little room for reaching out. Not only were tradeswomen often isolated one from the other, they were sometimes even ignorant of each other's existence. In various trades and in many cities, shadow training programs existed whose participants were primarily workers of color; they were called trainees, not apprentices. Kathy Walsh had been amazed in her second year to meet the other female apprentice carpenters. But it was still later that she learned that there were black women being trained in a separate program in Kansas City. They did not work on union jobsites, but in an extended pre-apprenticeship-type program.

UNBEKNOWNST TO US AT THAT time, there were two or three other female carpenters who were going through a pre-apprenticeship in the Hometown plan, affirmative action plan that they had at that time. They had their classes totally separate from us. There were three black women in that program, which was associated with the builders but separate from our apprenticeship. They were called trainees.

I believe the official reason for the separation was that they were on assistance and were getting government funding for their training. They had their two years of pre-apprenticeship first, before they ever even got the chance to work their 600 hours to get into the union. They were doing hands-on training within their own program. They had their crews and what they would do is, they'd get an old house that needed to be fixed up.

As Kathy came to understand it, the trainees received a stipend, rather than union wages, and didn't begin building toward their union pensions until they began work, two years later, on union jobs. Nevertheless, they were being trained at the same school location.

THIS IS IN THE SAME building—two separate programs. Bizarre, huh? I don't know what the logistics were, but the five women that were in the regular apprenticeship program—and there were some black males in, too, just about as many, hardly any—we counted towards the numbers in this other pre-apprenticeship training.

But we got to go to the real one.

In New York City, trainee carpenters studied in the same classrooms as apprentices like Barbara Trees and did work on regular union jobsites. Although Barbara sat beside them in class, their situation and status were unclear to her.

I NEVER COULD FIGURE IT OUT. They would be a trainee and they'd get work, but they would not be asked to join locals. Through their whole apprenticeships. And sometimes those same people had extended apprenticeships, which to me made no sense. It was kind of outrageous. But it seemed to be a racial thing, that the people of color were the ones who were chosen to be trainees and were not asked to join locals.

My feeling is that if you put people of color in a special category—when you've got an apprenticeship system set up that's just fine—you're saying something is wrong with those people. Some people, including myself, went into the carpenters union pretty ignorant. I didn't know how to read a ruler. I learned that in school, okay? Somebody else could learn it as well. I think what they were trying to say is, we don't want to just let them in, we're going to have to make them into a special category. From what I can see, they've suffered from it. The only reason I was not in there is because I'm white, I believe is why. As far as I know there were no white women trainees. The black women who were trainees, they were not asked to join locals in seven days or thirty days like I was.

For a year and a half before she was allowed to become an apprentice, electrician Barbara Henry, who is African American, worked as a trainee on a union jobsite in the Washington, D.C., area. She entered the industry well prepared by a stipended pre-apprenticeship training run by Wider Opportunities for Women (WOW)—six months of classroom instruction and on-site experience. She had already gone through another work training program sponsored by the District government, rehabbing old houses. There she had been leader of the drywall crew, heavy work since "the organization I was with, they weren't using screw guns, so you were tapping it up with a hammer and nail." While being trained at WOW she was assisted in finding a placement in her chosen field, with an electrical contractor. She worked on the same job first as a trainee and then as an apprentice.

I WAS SENT OUT THERE FROM the ground up. I mean, it was nothing but mud. I stayed there from when the job started until it completed. That was from '81 until '84. So I got to see everything.

Before I even was hired by the company, I had to go down to the union and get an apprenticeship license saying that I was going to be an electrician and all that. They gave me a list of tools I needed to have, and then they sent me out.

Two years in a row she tried to get accepted as an apprentice.

SCHOOL STARTS IN SEPTEMBER, AND I didn't get employed until, I guess, October, November, so it was a little bit too late. I'm believing that was the reason why I didn't get accepted at that time. I applied again the next year, which was in '82, and I guess because I was a tad bit late for the interview—I mean, fifteen minutes, I believe—I didn't get in then.

At that time Barbara was aware of very few African Americans in the apprenticeship program in Washington, and no whites in the trainee program. Whites who didn't make it into the apprenticeship often went into an "R" program, working on certain designated sites at a lower, "residential" rate. Regular apprentices went to school two nights a week; R apprentices, one night a week.

A TRAINEE, BACK THEN, YOU DIDN'T go to school. And that's where they sent the majority of the blacks. Because they was, I guess, segregated at one time.

I figured it was just work, and I believed in the back of my mind that I would become an apprentice eventually. I knew I did not want to be an R, because in the long run you wouldn't be able to make the same amount of money and you wouldn't have the same clout or prestige, okay. Because you go on a job and they say, "Oh he's an R worker," and you would see the whole demeanor would change totally.

As a trainee, I did all the grunt work—digging ditches, moving trash, unloading tractor trailers for days in and days out. You didn't use your tools that much. I had to clean out the trailers, about three or four of them—it was like a little baby office. I'd have to go in there once a week and wash and clean out their toilets and mop their floors. I felt that it had nothing to do with electrical work, but the electricians would tell me, this is the way you step up.

Our director, Mr. Larry Greenhill, advised that I go take a math class at a trade school. He suggested that—since I had been applying since '81—that this would further along the Board that I was really interested in this. So I went to school at night. He was trying to help us steer because, like I said, they would put the whites here, the blacks there. When I came in, there was only one black woman was even in the apprenticeship school.

On her third try, Barbara was accepted as an apprentice.

IN '83, I PROVED TO them that I was persistent. I went down and I told them that if I did not get in today, this year, I was going to go into the R. I got into apprenticeship school and on the job, it seemed like everything changed.

Folks started teaching me. It was like, Well, Barbara, let me show you this. Because while you're a trainee you only did the grunt work. I believe they figured a trainee wasn't really nothing but labor. It gave me, I guess the backbone or the guts to say, Look, I've done this grunt here now for a year and a half. I want to learn something, I want to know the *whys*. I would still do what they told me to do. But any opportunity they gave me to go watch somebody else after I did what I was supposed to do, I would go over there.

You got to know *why* to dig the ditch—the wall is here and the pipe's supposed to be in that wall and the pipe is over here. Before, the man said, just dig it from here to here. And an apprenticeship would come and bend the pipe and get it up in there.

Barbara was never exactly certain how the apprenticeship opportunity became available, but there was a significant group of women that came in alongside her.

I BELIEVE BACK THEN, SOMEHOW OR another, a door opened. I don't know if it's true or not, but I heard that this black guy had won a suit, and instead of him being paid, he requested that you had to accept so many blacks, male and females, into the program. Instead of him getting money, he made them do it that way. I believe that's how a lot of us got into the door at the time, if that is true. All of a sudden, we had women that was just—whew—coming in. Because when I came in, we had about eight women that came in at the same time, two white women and the rest were black.

CHAPTER TWELVE

Ceilings and Access Panels: Economics

> And then more women started coming in the painting field, and apprentices too, so they'd rather hire a 50 percenter than a full journeyman.
>
> —*Deb Williams, Boston*

No matter how much a tradeswoman might enjoy the craft or camaraderie of her trade, it was a job, and had to meet her financial needs. Economic considerations were pivotal to women's decisions to stay in or leave union construction work. Their hourly wage was high and, by union contract, equal to men's—but their annual earnings depended on how quickly they moved through the apprenticeship pay scales to full journeylevel rate, and how much of the year they actually worked.

For ironworker Randy Loomans in Seattle work was steady:

IT'S GOOD MONEY. I'VE MADE $44,000 last year with a month off. Now that's a good year, wouldn't you say? I don't even know very many men that make that much money, let alone an uneducated woman.

Construction work is by nature temporary. Availability of jobs not only rises and falls during different seasons of the year, but varies from one year to the next, region to region. Boom years when the skyline is punctuated with construction cranes and the emerging steel of hotels, office buildings, and hospitals can be followed by lean years when unemployment is high and tradesworkers must find other work locally or travel to another part of the country for union construction jobs. Diana Suckiel's concerns as a plumber in Kansas City could be voiced by anyone working in the trades:

IF I KNOW THE WORK is good—when you know you're going to have employment at least nine months out of every year, if not more—I'm more apt to want to stay in the trade. Like Bartle Hall will almost be two years for me when it's completed, and that's good, because you could make plans. You can budget your money and you can get things you need to help tide you over when there isn't any work.

The times that I'm laid off for a very long period of time, that's when I start thinking, Boy, I don't know, I ought to find something else, this is ridiculous. And then when I come home at night and I'm too tired to do anything else and my body hurts, I'm thinking, Wait a minute, how much longer can I physically do this?

When speaking to women considering a career in the trades, Barbara Henry is candid about the economic ups and downs:

I TELL THEM THAT IT'S NOT always gravy. Just like myself now, I'm out of work. I believe I happened to be at the wrong place at the wrong time, because I know from '81 up until '89, I worked steady. I was getting so much overtime, it was unbelievable. I was making almost sixty grand a year and then it's like, the bottom fell out. If you're ready to go on a rollercoaster ride from time to time . . . sometimes it's a steady uphill climb and then, you're just going down. If you can handle that, then it's something interesting for you.

Once you have it, they can never take it away.

At best, when work in a local is divided fairly among members, one needs to plan for that financial rollercoaster. Cheryl Camp, one of the few female electricians in her local in Cleveland, has found her trade a solid economic choice for her family, although less lucrative than those unfamiliar with the industry's swings might expect.

I JUST WENT BACK TO WORK this past August. I was laid off work eleven months. I don't think there's any preferential treatment. When the work gets low, we just get laid off, you know, there's no difference.

When people hear how much money you make per hour, they look at you with an incredible expression on their face. "You make what? Twenty-three dollars an hour? You're rich." No. I don't know where the money goes, but it's not there. My children are in private school and I own a home, so I have bills. I wouldn't be able to do the things that I'm doing working in any other kind of position, put it that way. But I also had imagined that I could have been much further ahead than I am now, too. But due to the unemployment and some other personal things, I'm just not where I thought I would have been.

Many tradeswomen, however, perceived a less than fair distribution of jobs. As an apprentice carpenter raising a nine-year-old and an eleven-year-old, MaryAnn Cloherty realized that not only was she not earning enough, but the inconsistency of the work created constant turmoil in childcare arrangements. She was laid off from her first job after three or four months.

T HE JOB WAS CERTAINLY NOT finished. I got laid off, I'm sure, because they got the hours that they needed, female hours. I was out of work for several months. And then I got sent to another federally funded job. That was down in Rockland at the sewage treatment plant. It went on like that. I'd get a job, it would last two or three months, I'd get laid off. I'd be out of work two, three, or four months. During those periods of being laid off, I'd go back on welfare.

It was no better than being on welfare, in the long run. Plus, I still had the problem of who was watching my kids. I had to be on the job at seven. I'd have to get teenage girls in the neighborhood, but they weren't all that reliable. Sometimes they wouldn't show up. And then my neighbor would tell me she saw a party going on. Or the kids would complain. It was very stressful.

When I would go to school—our apprenticeship school was at night—the guys would be talking about the different jobs that they'd be on, and I remember being very envious of them because they were doing things like putting windows in and putting roofs on. They were talking about side work they were doing. It was a whole network they had that I didn't.

I think I felt like I didn't have any allies whatsoever. I didn't find anybody approachable, and I really got the sense that the only work that I would get would be work where there was some type of compliance going on, and that's when I would get a call.

I started looking for other work. I found another job in a completely different field. I gave them two weeks' notice. Three days before I was leaving, I was kind of joking around. I was saying, "You only have two more days to take me out to lunch." They said, "You should know, we found a replacement for you." And I remember being struck by what they said: "It isn't a woman." I was struck by the fact that they referred to me as a woman. I thought that was a big advance—but then they finished it with, "It's a nigger." To me that really summed up why I was leaving. Every day that I went to that job I felt like I had to put armor on. I would just come home wiped out. Not physically—because in fact they weren't demanding that much out of me—but emotionally I was just wiped out. Then there was a whole other component of my kids. I couldn't really guarantee them a stable home life.

I just felt too alone. And I stopped thinking it would change and that anything would ever get better. I told the supervisor I was leaving the job. I never dealt with the union. They told me not to come to union meetings, so I really didn't feel like I was a member.

A month later I got a phone call from the business agent at the hall, and he said, "So, what are you—in the union or are you out of the union?" "No, I'm not staying in the union." He said, "Well, I'm going to send you a formal letter. Would you sign it and return it to me?" I said, "Fine." He said—I remember it really haunted me—"You fought so hard to get in and now you're leaving so quickly." It was like, I don't want to get into this with you. It was a form letter that I agreed that I was leaving the apprenticeship program. It wasn't personal at all. They made it very clear that they didn't want me there.

Yvonne Valles entered the painters union in Los Angeles with clear financial goals. Not only did she apply herself to the trade, she thought strategically, even switching locals to improve her work prospects.

A LL THE BIG WORK HERE in L.A. is union, from the hospitals to the schools to the utility buildings and the high-rises. It was in the '80s when they started hiring women, when the unions were pressured. They wouldn't get city, state, and federal contracts unless they hired women and minorities.

You'd find out which contractors were the ones that would hire women. Even the apprenticeship school knew, because they'd make the contacts for us while you're going through your apprenticeship. They used to call the same contractors over and over again that would hire women. And they keep us for maybe a month or two, maybe a week. You never knew exactly how long you were going to stay.

Most of the [male] apprentices that were just as good as I was, they had work all the time. They'd get in with a company and they'd stay there. They'd get to work with some of the better contractors that did all the really nice jobs. The contractors that would bid on the big jobs like renovating an entire hotel, that meant months of work. There was one company, two companies I worked for that kept me on for a long time.

I was in three locals, I changed locals. The first local, the business agent there—you talk about an old Mexican macho! He just felt that women belonged at home making tortillas. So he was real hostile with me. The reason I got into his local was because they got all the work downtown. They did a lot of paperhanging, and I wanted the paperhanging jobs. I wanted to learn painting, too, but wall coverings—if you became a journeylevel person you would make top pay, you would make $21 an hour. With painting it could vary. New construction pay was $21 an hour, but maintenance pay was $18 an hour.

But then after I experienced the hostility and resentment, the head of the apprenticeship school said, Why don't you try the Hollywood local. They have a lot of paperhanging jobs that they get calls for over there too. So I got into the Hollywood local and that was a big mistake, because I never got a paper-hanging job out of that union.

They sent me to work for Mario's Painting. Steady, steady work. And I mean, they were impressed with me from the beginning. I got a lot better treatment there. And I worked harder than most of the guys too, you know? I wasn't really trying to prove anything to them. That's just me. When I like my work I get really involved in it. But I stayed with them for about a year and a half. When I was getting closer to becoming journeylevel, then they let me go. That's another game they play with the women.

You know, there's nine levels you work up. When you're ninth level, that's your last pay scale. It can take from four to five years. Some of the guys only go in for two or three years. Because if they have family or friends, they'll start them at fifth level apprentice. You know, their father or uncle or somebody that owns a painting company or that knows somebody in the trade, they'll lie for them and say, hey, he's got experience. So even if these guys don't even know what they're doing, even if they can't hold a paintbrush, they'll start them at fifth period apprentice when you already can get benefits.

And they'd get overtime. When I'd find out somebody was asked to work overtime I'd go and confront the superintendent or the foreman. I'd say, How come you didn't ask me? They'd always have some excuse. They'd get asked to do side jobs too, make extra cash. When they'd need a couple of people to help them out they'd never ask me, and they knew that I was a better worker. They don't want to see women make money. For some reason it really bothers them—why should she be making as much as me?

One thing I liked about Mario's Painting—they gave me more independence. But when I became too high-paying then they let me go. And then what really, really hurt me—because I thought I was going to stay with them because all the guys liked me and they were always requesting me to work with them on their crews because they knew I was a hard worker—anyway, I just went and bought a brand new truck—and they laid me off a week later. They knew I had a new truck.

It was money. They just didn't want to pay me $19 an hour.

I only worked for about a year journeylevel, and then I couldn't get work anymore. The guys, they told me this from the beginning—once you become journeylevel, it's going to be harder for you to get work, because the contractors want to keep the older guys, the people they've had around a long time, rather than keep a brand new journeylevel painter, and especially because you're a woman. It was like a curse, but it was true. The only work I could get was maintenance work.

Anyway, work started slowing down in 1990. I was off work from December through April of 1991. I'd go to the hall every week and sign up. I was number one on the list and I couldn't get any work. I'd get work for maybe a week. I knew that there's dozens of jobs throughout the city where they need to have women. The Community Redevelopment Agency which awards these contracts to the painting contractors, downtown L.A., they had all these projects going and yet they weren't hiring women. And when the CRA would say, Where are your minorities and women? the contractors would say, Well, no women ever apply.

So I finally called the Community Redevelopment Agency and talked to two of the head people there in charge of the contracts and I said, What's going on? I've been off work now going on three, four months. The union is saying there's no work. I know you need women on these jobs downtown. She says, Well, call this contractor, that contractor, they're supposed to hire women. I'd call and the contractors wouldn't get back with me, they refused to talk to me, or they'd say, We're not hiring. I'd say, Well, I know you need women, so hey, I'm here, I can use some work. Well, we're not hiring right now, we already hired some women. I said, Well, who are they? How many women? So I called the CRA and I'd say, Hey, I called these people that you told me about and they told me that they already have women and they're lying, because there's not that many women painters.

They wouldn't do anything. I don't think they wanted to hear it. And then I went to the union hall one Monday morning—we'd go every Monday to sign up if we were still out of work—and my business agent said, Oh, by the way, Yvonne, there was an ad in the *L.A. Times*. They need a painter at the state prison in Tehachapi, you should check it out.

Yvonne had already been considering a switch to civil service and had put in an application to the city. She took the advice, and applied for and was offered the state prison job.

IN A WAY I WANTED to stay with the union because with the union I'd be working for more money. But then I was starting to balance it out and I thought, well, at least with the state I'll be working twelve months out of the year, I won't lose my benefits like I do. When I'm off work at least two months or three months with the union I lose my benefits. You have to have so many hours a quarter to be eligible for your medical coverage. So I decided I was going to take the job with the state.

Tradeswomen who felt that discrimination limited their earning ability in construction considered marketing their skills elsewhere, often where hiring and advancement decisions were made by less subjective standards. When tradeswomen worked steadily as low-rate apprentices but not at full journeylevel scale, these high-wage "nontraditional" jobs essentially reproduced the same inequity they were meant to address: that jobs primarily held by women tend to be lower paid. Since the goals for women on federally funded construction projects were never increased beyond the "initial" 6.9 percent established for 1981, there was no incentive for contractors to hire women at full rate. The entrance door for many women became a revolving one.

On the job where she turned out, going from apprentice to journeylevel, Gloria Flowers experienced physical and verbal harassment. In a pattern familiar to women who have experienced harassment in any industry, a layoff followed.

IT WAS SCARY. THAT WAS the beginning of my economic recession. I was laid off almost a year and a half. I worked hand to mouth, did some side jobs for friends, I went back and I rehoned my secretarial skills, took a computer course and brushed up on my typing skills. But it was a depressed period for me because economically, the money wasn't coming in. I was like eight months behind on my rent, my gas bill was over a thousand dollars, I was about ready to lose my car, my credit went right down the toilet. Just before I bought this house I got myself straightened out.

The local is mostly Irish extraction, okay? And you have to be kind of related or have some contacts, or be quote unquote a good union member, to kind of keep yourself going. That just seemed to be the way it was, that kind of a clannish set-up. If you weren't somebody's cousin or brother or something, you just didn't get the jobs. You stayed out of work longer.

Gloria enjoyed many aspects of plumbing, and maneuvered to stay in the trade. While unemployed as a plumber, she got work with the Pipefitters local, and applied for certification as an inspector.

THE NEGATIVES WERE THE LONG periods of drought and unemployment. I took the best advice I think I got from anybody, from the instructor who was the apprenticeship teacher at the time. He said to learn all we could and to try to expand our base, because economic times are hard. 'Cause the late '70s and early '80s were bad for construction. All the way up until '85, '86 they were bad. I listened to him when he was talking about the different avenues that plumbing could take you into. I went down to the state in 1986 and took the state exam for plumbing inspectors, hoping that that would be another avenue I could pursue. I passed the exam in 1986.

A lot of guys said, You're crazy. I said, Hey, I want longevity. I want something that's going to take me into retirement, you know?

Somebody was saying, "We want you to talk to other women about getting in the trades." But I says, It's naive to go out here and blow air at these women that are coming in. I consider myself fortunate, in that I know that being in the trades has improved my financial condition now—because as an inspector I'm making a good wage. But if it hadn't been for a lot of self-motivation in going forward, I wouldn't have gotten a lot. Some of it's due to them, but a lot of it is due to me also is what I'm trying to say. It wasn't until I started pushing and pushing that things got better for me. Because I didn't really have that much help, per se, from the local as far as keeping employment going.

That's what I mean when I say I wouldn't, in clear conscience, go out here and encourage a lot of women, because, they can go through an apprentice-ship and serve their time and get out and the work not be there for them. And they're back to square one.

There's a lot of discrimination, and when it comes to the fact that if there's only 200 jobs available and they got 200 relatives, they're going to put their 200 relatives to work first. I think the thing that has kept minorities and women going are the federal set-asides. And if it wasn't for that, we wouldn't be working, you know?

Paulette Jourdan left the union sector to become a self-employed tradeswoman out of economic necessity. Although as an apprentice her longest layoff was two months, as a journeylevel plumber, she found herself mostly unemployed, despite an abundance of federally funded jobs in the Bay Area. She attributed her lack of work to the fact that she had accused the dispatcher of unfairly putting another name before hers.

I REALLY DIDN'T UNDERSTAND THE POLITICS. There was no way I could have at that time. The politics say, If you are not who we want you to be, and you don't act the way we want you to, you're not going to play. And they mean it. And they have ways of keeping you from playing the game. They forced me to leave by not letting me work. So I finally did get it.

I made a lot of money when I was an apprentice. I mean, I didn't get rich, but you work more consistently when you're being trained. I stayed with them two years after I graduated, which wasn't much, because I only worked three months the first year and four months the second year, and that was it. I couldn't take it.

Different companies will hire you, but the union always dispatched you. The jobs were always very short. I'd already started moonlighting. Whenever they'd lay me off I'd just find jobs around town.

When you're an apprentice you're competing with, at the most, sixty other people, because there are only so many apprentices they can take in at once. But once you graduate, you're competing against 600 people who have ten times as much experience as you do.

It's fifty percent name call, or it was when I left. Fifty percent name call means, for every one they call by name, they have to take one off the list. I knew I wasn't ever going to get name called, and I'd be at the bottom of the list, maybe 89 or something. It would take months to get to the top. And then when you get close to the top—you get to number ten or something—the dispatcher in the office—she hated me—she would call me up with a one-day job or a two-week job. I think once she called me to go to Oregon for three weeks or a month or something.

I never got a job once I graduated that lasted more than two months. If you work three weeks consistently you go to the bottom after the third week. The crowning touch—just before I quit—they put me on a job that was supposed to last three weeks. It got down to like the day or two before, and I called—because I didn't want to do it underhandedly—I said, I want to ask this company to lay me off, because I don't want to go to the bottom of the list. And they said, If you do, I'm going to put you there anyway.

They kept me three weeks and two days. It was like, you don't really know what the forces are at work, but you know there's no real way you can deal with it. I said, forget it. I can't eat like this. So I left.

They were fighting it tooth and nail. Now, there were lots of federal jobs going on and not very many women, and I only worked on two federal jobs in six years? There were only two black women in the whole Alameda County Plumbers, and neither one of us were on too many federal jobs.

What I discovered, since I left the union and have been forced to do this work on my own, is that I have this little part of me that likes to sort of feel like a rescuer. Okay, you don't charge as much as the guys would charge, and you do as good or better job, and you go take care of people's emergencies sometimes. And people compliment you for not being like the guys—for not charging 'em seventy bucks an hour or for not messing up their house and leaving a mess when you leave. It's a kind of acceptance that's sort of a bonus to doing the work. If I was a better businessperson I could probably make better money. But basically it's good, solid, honest work and pays good. People respect you.

Hoping to produce better enforcement of affirmative action standards and fairer hiring practices, tradeswomen looked for pressure points. Irene Soloway described efforts by United Tradeswomen to have women hired on a major project in New York City.

I GUESS THE MOST POLITICAL THING we did was when the convention center, the Javits Center, was being built. The Urban Development Corporation was overseeing the whole project, and we tried to meet with them around enforcement. Then they hired some women laborers.

And then they laid off the women laborers, and we objected to them laying off the women.

We had a demonstration outside their offices in Midtown on 42nd Street. It was not just women, it was men, and it was men from the minority coalitions. It was a lunchtime thing, altogether maybe 100 people. It was enough publicity to have them put the women back. We had all these women up in their offices, you know, and we negotiated around a table. We told them we wanted twenty people to come to the negotiations, and it was all tradeswomen negotiating.

We had them in a good position because we had press and they didn't want bad press on the convention center. It was like right in the public eye. And then, of course, as these things always happen, the women who got back on the job never paid their dues to United Tradeswomen and they never came to another meeting. We were real mad at them. So that was discouraging, but it was a good experience in terms of using power.

Irene Soloway at a demonstration organized by United Tradeswomen for the hiring of women on construction of the Javits Center in New York City.

Career decisions are ultimately personal, even for activists. After almost a decade in construction, Irene moved into a civil service job.

WORKING OUT OF THE LOCAL was hard to do. I couldn't speak freely about anything because the word would always get back to the local. It was real easy to be blackballed. I spent months sitting on the bench and there was no recourse. I got tired of that. I mean, I would say something about the business agent and somebody would say to me at a meeting, We heard that you're talking about X, and don't do that. I just couldn't stand that after a period of time—being that cowed. I just didn't feel like I was going to win any battles over there and so I went with the City, I took that route.

The numbers [of women] stayed about the same. You know, they're not permanent jobs, people get laid off all the time. It could take you forever to finish an apprenticeship if you're not getting consistent work. It just gets discouraging—the nature of the industry, not having strong advocacy, the individual women being tired of being out of work.

I've been working for the City for the past six years and I'm the only woman. There's a different work agreement and the wages are always about three years behind. Actually, now happens to be a period where there have been layoffs, and jobs that really turn out not to be as permanent as we thought. But, in the best of times, I think civil service has a lot more potential for women. It's not perfect at all, but there is a merit exam to be a supervisor of mechanics. You work with a group of people consistently. I've established myself once, and now I don't have to prove myself all the time, so I can be human. There are very few women in civil service because you have to have five years of journeylevel experience.

Carpenter Lorraine Bertosa also perceived a lack steady work, caused by discrimination, as a primary reason women leave the trades. Along with other members of Hard Hatted Women of Cleveland, a grassroots organization of women in nontraditional jobs, she worked to leverage enforcement of affirmative action standards on the city's larger construction projects. On the Gateway Stadium job, a $400 million project on a 28½ acre site, their on-site monitoring brought female participation to 6 percent. Still, for Lorraine, getting work has largely depended on personal contacts.

I THINK GUYS ARE OUT OF work, but—there's none of the women working at times. And that's real off in terms of the percent. I've had to fight to get on some jobs. The Sohio job I literally went down there eight times or something to get on. Finally, somebody saw me and knew me and told somebody to hire me. It's the only way I got on. The job was loaded, you've got a thousand guys working down there—and you don't have any women? Give me a break. Yeah. It's still like that.

If you can't get a job, or if you have to fight so hard all the time just to get on the goddamn site—who needs that? The guys are getting their buddies on. I got on a lot of jobs because my buddies were on, too. Hard Hatted Women tried to put pressure on the Sohio building, but I couldn't get on there for love or pete until somebody saw me and knew me.

Where's the push to hire the women? If the push had stayed on we could have winnowed out the women who just wanted to make the buck and gotten the average woman that was willing to do the work. There's no steadiness with the work. I couldn't believe Gateway. I got really pissed off. I went to this meeting with these head guys you see on television all the time about this Gateway project. And their EEOC officer. I've never had any luck with EEOC officers. They just talk you in circles. We're sitting there at this table and he starts saying something about, "Well, you know, where are we gonna get these trained women?" I just looked at him and I said, "It isn't about any of that. Can you provide a job? That's it. You got a job, we can get you the women left and right. Is there a job?"

They're trying to push it back into "there's no women." *There's no opportunity.* There's no active opportunity, like there was with Carter. The only active opportunity that still held was with our governor before this one, [Richard] Celeste. His wife was pushing, so if it had state money there was still somewhat of a push involved. And they found new ways to get around it. The latest thing they're doing—they did this to me on this winter project—is to have you work for this short period of time and then they lay you off. You're on the books as being in there. They don't check the books out. You could have been laid off the first week. But your name's in there.

I think it's union-based dragging their feet. They want to keep their constituency working and it doesn't involve women. It's really sad. They want our support. But they're not supporting jobs for women.

The informal hiring networks all tradeworkers depend on in a fluctuating industry are less available to women. In Los Angeles, ironworker Mary Michels, with strong memories of her early years in the trades when affirmative action enforcement was expected, tried to activate those responsible for monitoring.

THERE IS A LOT OF companies out there that just don't hire women, period. You hear about affirmative action. To me, that's a total joke. There is no affirmative action as far as I'm concerned.

The convention center down in L.A.—that's really interesting. That went on for years. One female was out there for maybe a month. They had one woman out there, and they had hundreds of ironworkers. It just infuriated me—a big project like that and there's no women. I did try quite a few times to get on it but couldn't get hired. I started checking into that one. And I got absolutely nowhere.

I went to talk to the head supervisor for contract compliance for the City of Los Angeles. He really wouldn't give me an answer. He was going to check into it and he was going to go by the federal compliance. They're supposed to go to an employer and, you know, pretty much force the issue. But they don't.

I went half, just as a citizen, because I was very angry about it. And then I went with Anne Brophy and we went as Southern California Tradeswomen. He sat there twice very nicely and listened to us for an hour and a half, and then he sent a woman to one of our meetings. Did they do anything? No. They don't even keep records. They say they don't even have statistics on the hours women work out there. They only have it on a weekly basis craft by craft. So if they wanted to know how many women, they would have to go through each week and look. They don't have it organized. That's disinterested.

There's no enforcement of affirmative action that I know of except the Century Freeway and now the L.A. Board of Education. But the federal compliance doesn't do any affirmative action. They had no female ironworkers on any of their federal jobs, so I don't know what those people do. I don't have any use for them.

Even the County doesn't do anything either, because I called up the County about it and they told me to go down to the Fair Employment and Housing Department and I would have to make a complaint. They don't check to see if these companies have women and why they don't. They only take care of complaints as they come in.

There was a lot more women in the trades in the early '80s. And they were working all the time. There was acceptance, the guys knew, Hey, the laws are changing, we got to get used to it, women are getting into the field. There was an attitude change, absolutely. And then somewhere along the line, it went down the drain. The late '80s it started hitting me. We have all these women dropping out.

I thought there would have been loads of women by this time. A fifth— even fifteen percent, that would be great. A few women on every job, that would really be nice. There's no doubt that the women can do it. I just figured it would be more of an acceptable thing and they would get hired. They're not being hired.

You go down the union hall and you can bid on your jobs and I think that's fine. But that's just a very small part of it. You get jobs by the people you know. The males can go out and drink together. They can call each other up on the phone and stuff like that. I have a very limited access to phone numbers. In fact, I've got a guy that lives five minutes from here, he doesn't car-pool with me because his wife doesn't want him to. Now, what I'm going to do to that man, I have no idea. Guys don't want to give me their number because their wives just would not understand a woman calling up their home, asking for their husbands. They cannot relate that I'm looking for a job. I do have some numbers but, in general, no, it's not acceptable.

There was a guy I was working with, he tried to get me on several jobs, but he's told me, "You know, I've tried to get you and a couple of the other girls on and the minute I mention the name, they says, 'Lookit, we do not want to deal with females on this job. That's it.'" The foremen, they don't want the headaches, whatever that means.

It's kind of like integration in the early '60s. Blacks weren't even allowed in restaurants. They forced integration. You have to force the same thing with women in the trades, and then people are going to get used to it and then they're going to have attitude changes. But it's got to be forced or you're not going to get it.

For tradeswomen who did not move into civil service jobs or self-employment or make it into a company's core workforce, their long-term financial prospects are a major concerns. Ironworker Gay Wilkinson attributes women leaving the trades primarily to

LACK OF WORK, THE FACT that you're so underemployed. It's just very hard if you're looking down the road to a future to say, Gee, am I only going to be able—for the next ten years—to live on working three months out of the year, or twelve weeks, or five months? What I've seen happen now is women are employed less and less. The unions seem to feel more comfortable about stating, There's no pressure, I don't have to hire you.

So where has all of this brought us to? I mean, I'm 52 years old. I've got ten more years, and I've got to get ten years good time in order to have any money for retirement. And young women I know coming up have got to have some kind of assurance that there'll be jobs for them if they go through this training. And I don't see it happening. I see us becoming more and more frustrated over the fact that our numbers aren't increasing, and the feeling against us is getting stronger and stronger and more and more people accept it.

This is one of the pioneer issues: have we had any women who have had a full work life in the construction trades? Do we know anybody who started even at 30 and has worked thirty years in the trades and retired with a fairly decent work life? Is there anyone out there like that? Not that I know of. Is it actually going to be a thing that's going to happen? I think that's something you have to wait fifteen years down the road to see, if there's even one woman who has gone from apprenticeship to retirement and worked in her trade, where it's a totally accepted thing for guys.

CHAPTER THIRTEEN

Ceilings and Access Panels: Leadership

> The men that have really recognized me and made me foreman are men that are sure of themselves. And not threatened by a woman. They applaud having a woman, because there's a lot to say for us. I think we're a lot more dependable than men. More conscientious of everything we do.
>
> I think in time I will be running work. I think my time's coming, but it hasn't arrived. And it's taken a lot longer than it would have if I'd have been a man with my same attributes.
>
> —*Randy Loomans, Seattle*

In evaluating any career choice, the potential for advancement is always an important factor. For tradeswomen, moving into supervisory or leadership roles often required traversing minefields similar to those they had crossed when entering the industry.

One way in which tradespeople advance themselves is to take their craft knowledge into related occupations. Marge Wood considered it a normal career progression to follow her twelve years of experience as a plumber by working for the Wisconsin Technical College System, where she reviews curriculum for the classroom instruction the state requires of apprentices one day every two weeks.

I T'S VERY DIFFERENT FROM OTHER states in that ninety-four percent of the apprentices—and there's about 7,500 apprentices in Wisconsin right now—go to technical colleges for their required related instruction, instead of going to a union training school.

I applied for a civil service job to go to the state as a plumber. There were 150 people that took the test, I scored second out of 150, and I was hired by the University of Wisconsin, the second person hired after that number one guy, who was laid off. I definitely beat out people who had been my foremen on the various jobs. I did great on the test because I had just finished five years of school and it was a code-based test. I find out later that one of the guys who had been my foreman threatened to file a discrimination suit. He couldn't imagine that I'd actually gotten that job on my own.

I can't count how many times in my current job I hear, Oh, the State of Wisconsin, they steal all the journeylevel women and minorities. We train them and they pluck them away. The University of Wisconsin never called me up and said, Come to us! I applied for the job and I really felt that I got it fair and square.

People started saying accusingly to me, We invested all that time and training in you and now you've left us. And that really pisses me off. Because I got background, I got training, I got drive, I got all kinds of things. The same stuff in a guy would be showing initiative or being smart or being a credit to the trade. And for me it's like, You are a token woman and now you're gone, so we've got to train another one. How dare you do that! It's not worth investing the time in you. And of course these accusations without fail come from—I mean, these are daytime meetings—business agents who no longer work with the tools, and labor officials. And you say right to them, "Well, when is the last time you worked as an electrician?" And they will say, "Well, that's different." They totally do not get it. And yet when they go to promote the trades, they would promote it as, the trades is an opportunity to do all kinds of stuff, it's not just dirty work. You could be an architect, you could be a building inspector, you could do all of these jobs. But if you're a woman or minority and do that, you screw the trade out of the training they've invested in you.

In construction the positions of foreman, general foreman, or superintendent are desirable not only because they receive higher wages but because they offer intellectual challenge and reduced physical demands. As union members, the contract sets their pay rate, generally a percentage above journeylevel, and specifies the workforce size at which someone must be paid foreman's rate (and for the various positions above that). Careers differ. One person might become a company's "permanent" general foreman or superintendent, going from one job to the next and never again working with the tools; another might move into a supervisory role for a specific job and then be "broken down" back to journeylevel. But even a "working foreman," one with a small crew who still works with the tools spends part of the work week going over blueprints, attending job planning meetings, or ordering stock—easier on an aging body than hauling stock. Although she herself ran work a few times, the minimal advances made by tradeswomen surprised Kathy Walsh.

I THOUGHT SURELY BY NOW THAT we'd all be leaders in our field and there would be enough women that it wouldn't even be unusual to have a woman on the job. I can remember turning out, getting my journeylevel certificate, and thinking, now I've got it! I don't know why I thought it would be different. And going on the job and still not being the person who got to look at the prints, make any decisions, and be a lead person. I always felt like I was still a cub.

One job I worked on, it was a federal job out at an army ammunition manufacturer place here in Kansas. I knew the superintendent on the job, had worked for him before. It was a small job and a small crew. There were maybe eight of us, like five carpenters and three laborers and a foreman and superintendent on the job. The foreman was going to be gone for a couple of weeks at a different job and the superintendent was going to be gone for like a week. I was journeyperson and I was not the first carpenter hired out there. And he made me lead person while he was gone.

It was like *Mutiny on the Bounty.* I mean, I had been working with these guys for months, everybody getting along fine, doing a wonderful job. It was an easy job, it was a gravy job, we didn't have to bust ass or anything like that. By God, these guys—if I told them to do something, they'd do their best to find some way not to do it. I'd have to either get in their face and say, You do that now, or cajole them, Come on, you guys, we got to do this. I handled it. I think I was probably the smartest and the most competent person out there. It was form work, and I'd done a lot of it and I knew what I was doing.

Despite the many potential positives in the role, it was not uncommon for a tradeswoman to turn down a foremanship offer, as Cheryl Camp did when she was asked to be a sub-foreman at Tower City in Cleveland.

I REALLY THINK THAT IT'S LIKE the highest compliment within our union that you can get, for someone to come and ask you to be a foreman. They're saying, that we have confidence in you and we know you can do this.

I was in a fire alarm crew there and I was asked to become a sub-foreman. Our foreman on the job just wanted to designate a few people as his key people, that he knew that he could count on to handle the job, so I was asked. I was going through some personal problems then. I told him at that time, Thank you very much for giving me the opportunity to become a foreman—and basically the work that I did for him could have been classified as sub-foreman work—but I couldn't take that responsibility of running people at that time. I had too many other problems I was going through.

In our local it really is difficult for a minority to get into a position as a steward, or a foreman. You have to really have someone on the inside, to bring you along. Like when I was asked to be a sub-foreman for him, we had a great rapport. But he's a special person, too, basically going out on a limb. Because I would have been the first woman foreman in the local.

You have to weigh out, given the opportunity to become a foreman, how beneficial is it going to be for you to take that position. Because you're going to have personalities working for you that are going to try to sabotage your jobs. I saw it happen on a job where it was a minority foreman and he had his own area to do. He was given drawings and a crew. And the crew that he was given to do the work—because you can't select the people that work for you—it was like a pattern set up for failure. But miraculously, it went off pretty well. He made money for the company with what he was doing, 'cause he ended up getting some of his friends, that he knew personally, to come and help him out. They actually left another company that they were working for and came to work for him so he could pull the job off. It was obvious—and I could see it—the stress that he was under. He was ready to quit the company, period. Not just quit the foremanship position, but quit the company, 'cause, knowing that this is the mentality that they had, that they were actually setting him up in this position for failure.

Lorraine Bertosa's first response to being asked to be foreman was also to turn it down, even though the GF was a carpenter she admired and trusted. It was her first chance to work on a residential job since joining the union. She did end up running a crew of up to six, putting up interior walls for about seven months.

I WAS REAL SCARED TO DO it. I was out there the second day or something and Bob was just starting to pop the crews together and he says, "I want you to do this, Lorraine." I says, "Oh, Bob, I haven't even built houses yet. Let me be here for awhile." He says, "Nah, I think you can handle it. You start tomorrow."

He set it up really well. He had one guy, he said, "Now if Lorraine has any questions, I want you to answer them for her. She's got any question at all. That's your job." A guy on my crew. He was my age but he was real experienced at what we were doing. It was real cute. One day it was raining and the outside crew couldn't work. There was about six guys and we brought 'em in and we split our crews together. This guy who was helping me out, he took half and I took the other half. We decided to see who could put up the most walls in a day. We had a ball. 'Cause I was really pressed. I was gonna do this, you know, beat them. Billy'd come over—he was always smoking cigars—he'd come over and, "Oh, we already got that wall up, Lorraine." He'd do this all day long. But it was just a delight. We just pushed the snot out of each other. I don't know anybody else I could have ever done that with. That to me is the trade.

Bob would come and show me how much my crew was doing every week, "Well, you're making money this week, Lorraine." There weren't any women on the job. It would have made it even better for me if there had been a woman out there.

I lost that job 'cause finally some other work was starting up, some new concrete work, so they were starting to thin out the crews. None of the foremen were going, just they were thinning the crews down. The word came down that they wanted me to go to this concrete job. I was heartsick. I wanted to *stay* where I was. I wanted to get this skill. I couldn't understand what was going on.

Bob, my general foreman, *and* the superintendent of the jobsite both went to talk to the guy who kind of ran the office. Both spoke to him and said, "Leave her there. She's doing well." When these guys couldn't do it, then I finally went. This guy had been kind to me many times. When I had struggles with my parents he had helped me out and he'd done a number of things for me, so I felt okay talking to him. He's giving me this bullshit, "Well, we need rounded carpenters." Shit, I had done so much concrete work, I didn't need that experience at all. Later on it came through that he just couldn't picture a woman running a crew—it came sideways from the office. It was a way to pull me off, that was it.

I quit. I went down to the concrete job for one day. I was so heartsick, I quit. I wrote this incredible letter. I sent one to his boss, I sent one to him, I sent one to my general foreman. At that point I didn't realize he was so anti-a-woman-running-a-crew. I said, When somebody can't persuade you to reconsider something like this, that's so important—!

The strong affirmative action enforcement in King County and Washington State made the Seattle area a relatively favorable one for women to move into leadership roles both with contractors and in their unions. Working for a contractor newly organized into the union, Diane Maurer became foreman and then general foreman on two multi-million dollar projects. She first moved up when her foreman was laid off.

WE WERE DOING THE METRO Tunnel project, which was a pretty large project, and we had close to a hundred electricians between a swing shift and a day shift. We were under the gun with some slabs, and he handed me the radio and said, You're it, tag. I was given a full-size crew in the middle of ongoing projects. I had some access to the prints, but nothing near what you need in order to decipher wall finishes and various specifications, especially on a project like that which was underground. There was a lot of looking at architectural and structural prints, which I'd never done before. Actually, the other foremen were supportive 90 percent of the time.

I always kind of thought that as a foreman, you could somehow make things better. And that's really not the case, because there's a lot of forces that are going on far ahead of the time you get set up as foreman that you have no control over. They [the company] had a real hard time getting material that you needed for projects and keeping enough tools in good working order to keep everybody moving. When you get all those factors working together, you really would exert a lot of extra energy in trying to make such an inefficient system run somewhat smoothly, just to keep the guys' attitudes at a decent level. It was the first big job that they were taking on, so the general assumption was, good luck.

I'd been a foreman down in the tunnel for a year and a half and the general foreman quit. I was the person who'd been in the foreman's position the longest, okay. So they needed to set up a new GF—and they bring a guy from another station up there that doesn't know the project. He's coming to me and asking me, What about this? What about that? I'm pretty much having to hold his hand and tell him everything. That was pretty hard to take. I guess the reason why I stayed was because they set up a swing shift crew for the summer and they put me in charge of it, so I didn't have to deal with the guy.

When Diane finished that job, she was transferred to a new five-story construction project at a museum, where the same man was GF.

HERE THEY'VE HAD THE PROJECT for six months, and they have not done anything in terms of planning, and it was a very complicated job. As a matter of fact, it was such a complicated job, the original project was awarded at twelve million and it came in, finished project, at twenty-four million. J—— gives me the sketch of the basement slab, and he's got all these lines just zinging all across, coming over to this center thing. He goes, "I don't know if you can do it this way, but here." This is supposed to be the telephone conduit, okay? He hadn't looked into the thickness of the slab, if you could have that many conduits converging on one point....

I was just a journeyman and I'm running pipe for outlets and he hadn't even planned if he was going to go overhead or underground. We were stubbing them out in the slab and we were stubbing them out in the ceiling, which was a waste of time and money, duplicating everything. I could see, well, somebody had to start planning where things were going to be, because they were getting ready to pour these walls. So they set me up as a foreman, gave me some other journeymen to do the work while I started doing the planning.

J—— was transferred to run another project and later left the shop. Diane was made general foreman at the museum.

A FTER I WAS THERE ABOUT a year the company finished their work in the tunnels. Then they were putting all their attention onto the museum, and things started to get weird. Up until then I'd had one project manager. All of a sudden, they decided that I needed two. Whenever I would plan anything or order anything or want more men and want more equipment, the second project manager would have to review it to see if I really needed it. The superintendent didn't trust me to finish that job, do you know what I'm saying? Not that I had screwed it up or anything, but the reality of the situation was, *he was going to have this woman general foreman finish this job*—and that's not what he wanted.

Then they tell me one day, they hired a guy out of the hall to be a foreman—they don't know his name, but he's really good. Well, it turned out it was this J—— guy! You know, why not just come out and talk about it instead of being sneaky? He wanted to keep me on the job because I knew the most about it, but he wanted his buddy in there.

The museum was also another job that was under scale. They were having problems paying their benefits—guys would not touch that job with a ten-foot stick. So I would call my friends. My husband was working down there, some of my best buddies were working down there. I wanted people that I could trust to do a good job and that could kind of ride this thing out, because it wasn't going to be easy. I think that was another thing that bothered them, too. There was too much cohesiveness between their supervision on the job and the crew—even though the general contractor praised me and the men and the job we were doing, constantly.

I really felt that their plan was to bring J—— down there. I'd be getting the 20 percent over scale, but they'd be looking past me and asking [him]—and they'd end up embarrassing me into quitting. So my radical bone got the better of me, and we had a big wobble over benefits and I ended up leaving. The guys had been having this problem with the benefits not being paid, vacation money directly deducted from their paychecks that they could not get, and they were upset. The company, they were like, five months in arrears.

In a labor dispute over the unpaid benefits, the workers felt they were locked out, while the company claimed the workers had quit. Although she believed that, had the general foreman been a man, the situation would have been resolved without any job loss, Diane and her crew chose to move on rather than fight to get their jobs back.

I BROUGHT IT OUT OF THE ground. We had the rough on. We'd completed all the slabs and all the concrete sheer walls and we were working on wall roughing on the third floor and running the risers in the mechanical shaft. The building was probably 40 percent complete when I left.

When I worked as a general foreman on the museum project, I put a lot of effort into thinking through short interval schedules. I would list the project to be done, the amount of manpower it would take, the equipment that they would need in terms of ladders or lifts or types of benders—if they needed the large hydraulic bender for 4-inch rigid pipe, or whatever they were going to need, equipment-wise—and projected time-frame. I would update this every other week, and I would turn it in to the general contractor so he knew what areas he had to have cleared for us so we could proceed. Because if you don't say, This is where we intend to be, you can't go up to him and say, How come you're not ready for us to start? I also wanted my bosses to know, so that when I called for men and requested equipment they'd understand why I needed it.

I don't know that a man in my position would have gone to that extent. Of all the foremen and general foremen that I worked with for that company, I don't know of a one that ever did any kind of planning like that. I felt that it was important that I do the best job that I could because I was the first woman to be a general foreman in this area. If the job was disorganized and in shambles, it had my name all over it.

And I think this is kind of particular with women, that, not only was it important to me, but my whole self-esteem was attached to being able to do that. Which adds a lot more pressure than the job itself. I don't think men have that kind of emotional attachment to what they're doing. I'm sure they get stressed out and they want to do a good job, but I don't think that their self-confidence is attached to it. I just really went—not only the extra mile but the extra ten. It really wore me out.

Diane worked for another company as project manager, starting and completing eleven projects in six months. But despite her strong background, she found that the career ladder that men could move along fairly naturally did not function as readily for a woman.

I HAVE HAD SOME VERY INTERESTING experiences since I left the museum because, to me, having been a foreman, general foreman, project manager are pretty decent credentials. They'd look good on a man—they would, don't you think? Here I am, a woman with that background, and here's a man with even less of a background—they're going to talk to him and not even bother with me. I think there really is more resistance to women in management or supervisory-type positions because it's been a good old boys club for a very long time.

If somebody asks me to run a crew, I won't turn it down, because I'd like to do it again. I might be able to enjoy it more and even be better at it than I was.

A man chosen for leadership would likely be a consensus candidate or have the protection of family connections. A woman who pioneered in a position of responsibility, either with her contractor or with her union, more often had divided support and took on these new skills under the stress of heavy scrutiny. But women also had the opportunity to create positive structures. Karen Pollak moved her book to Seattle when the Army Corps of Engineers moved her husband's job there. Although she'd read about the job opportunities for tradeswomen in Seattle, she was still surprised.

WHEN I HAD LEFT KANSAS CITY I had just been offered a foreman job for the heavy highway company I was working for, which would have been the first female ever to have been a foreman in the state. But I said, "No, the husband's moving to Seattle."

In the *Tradeswomen* magazine in 1985, in the back of the magazine, there were all these different ads for jobs. Seattle Center and Seattle City Light were actively recruiting women and minorities in the trades. I came from the Midwest—they didn't give a shit about women and minorities. There was a big push out here, though. The unions couldn't get enough women. King County was looking for them. I was flabbergasted when I came to Seattle and I joined the local there and went to my first union meeting and saw—I mean, there was a whole row of women!

I was working for a woman contractor, woman-owned company. I worked for her for probably five years in Seattle. I became her foreman, her superintendent. She would farm me out a lot of times to the general contractors because they needed woman hours. It would be, "Well, she is my foreman. She's not going unless you put her in that position." I was never asked if I wanted to go to those other jobs. It was a deal made and it was like, You're going.

I always said that if I got into those positions, I would never be the foreman that I always had. Anything that was requested always came from the fact that I was right next to them, doing the same thing. I never asked anybody to do something that I wasn't doing already. I don't treat the women and the men any differently. They all get a fair shake. I had a fifty/fifty crew on the I-90 floating bridge. I had women apprentices, I had women journeylevel, I had women laborers. I had men in the same positions.

I made it a big point that I always was checking on the people's work. Because the one thing that I always hated was to do something, and the foreman would come by and look at it and never say a word to you—and then, come back when you're done and it's ready to be poured, and say, "It's fucking wrong, take it out, I want it done *this* way." That was *always* happening, and I never could understand why they would do that. Why waste time and materials? Oh, well—I realize that you want to make me look bad, so I'm fired and I'm down the road. . . . If you were going to be a good foreman, you wouldn't do those things. You would try to make the company money and you would try to keep your crew as happy as you possibly could.

I was a pretty lenient person. I didn't *have* to give coffee breaks, but I did. I gave two a day. If it was bad weather, I always made it a point that they got a few extra times to get into a dry shack to change gloves, get warm socks, little things that make working outside all day when it's miserable, tolerable. I bought supper if I ever asked any one of the people to work overtime that was past ten hours a day. I made it a point that *my* company would pay for their dinner—and we would go off the jobsite to eat. It wouldn't be like we would just sit there in the shack and I'd go get some deli sandwiches or something. We'd go get a hot meal and then come back.

I never got mad at anybody in front of anybody. If you're doing something and you're being really a shithead, "Go up to the shack for a little while. I'll be up there." Have a talk, and then they'd go back. But I didn't have to do that but just maybe once or twice.

Karen became one of two women on the executive board of her local, "the only local in the state of Washington, male-dominated, that has ever had two women on the executive board at the same time." She also ran for and won positions as delegate and executive officer of her Carpenters district council. As on jobsites, the responsibility to confront sexist and racist comments fell to her. At an executive board meeting:

L AST WEEK WAS THE LAST straw. I walk in. I sit down. "Hi. Hi. Hi." I ask for an agenda to the meeting. The man sitting across from me, which is the one that gives me a headache, says, "You're a fucking woman. You don't need a goddamn agenda. You don't know what it says anyway. And you couldn't read it if you wanted to. You guys should just stay home. Get out of our union." I stood up from the table. Everybody's going, "Don't hit him, because then he'll file charges on you and that's what he wants." It's like, I'm not going to hit him, but I don't have to stay in the same room as him. And I suggest I best not be counted absent at this meeting because of it. Because if I am, then I'm going to file a grievance. "Oh, well, I'm sorry. I was only joking, Karen." "Well, do you understand that that's not a funny joke? You've been told time and time again that your little comments about 'I'd like to give her CPR, mouth to mouth resuscitation,' that's getting real old." A union official's sitting back there going, "Why, you know how he is, Karen. We can't change him. Just ignore him."

I filed a grievance on him in our district council, because we're both district council delegates. He made a comment one night that—I had no other choice.

It was like, "You fucking squaws." Everybody's looking around and he's going, "Woo-woo-woo-woo-woo, woo-woo-woo-woo-woo."

I won that grievance. He got a fine of twenty-five bucks and was not to open his mouth at three district council meetings in a row or some bullshit. I was supposed to get a public apology at district council meeting. Well, I've never gotten my public apology yet.

Gay Wilkinson also found obstacles to women taking leadership, in her iron-workers local. The first woman to go through the entire stewards training class, she found that, although the education offered was excellent, both how the class was scheduled and how information was conveyed made it difficult for her to recommend.

I T'S ALWAYS AT A TIME when women can't make it, right? It's at dinnertime, it's on a Saturday afternoon, it's on some of the most outlandish days you've ever heard. And the whole class, from the time you get there till you go home, is the most degrading, belittling thing imaginable. Yet we're talking about a steward who is supposed to stand up for your rights.

The analogy they use is, if you as a man run into such-and-such a situation, you would have to be a person who sits down to pee not to stick up for the union. I'm sitting there, I said to him, "You mean to tell me, basically what you're saying here is, if a man has a problem with an issue, he would have to be a woman not to stand up for it?" I mean, to stand up there referring to everybody as cunts—when you're sitting there in the front row! It got so bad I stood up and walked out. The worst possible thing you could be is a woman. If you're non-union, you're like a woman. If you do anything disruptive to the union, you're like a woman. The analogy—that you have to sit down to pee, you have no balls—they're all analogies that point to women. It's like, we're the worst possible thing that can happen to unions, that can happen to men, that can happen to the industry—and this is steward training! I mean, where are people's heads? So one woman has gone through that entire class, from end to end. After that class I went to the executive board and I told them, "You people are leaving yourself wide open to lawsuits. This is harassment, this is sexist, it's bigoted, I mean—you name it, this is what your class is."

Stewards class goes from 6:00 to 9:00, five nights a week for three months. This is a long, entailed process. You do contract law, you do negotiating. These are all things that might have to be used on the job. You have to learn CPR, you have to be certified. It's an excellent opportunity for someone to learn about how a steward actually works, what the interaction is between the union and company and how problems are solved. And it would be good for women and minorities who have never had contact with unions to use this as a good educational tool. But it's done like you're in the goddamn barroom. It's that male kind of—if every other word I say isn't *fuck,* I haven't made a sentence.

These are the executives of our union who are going to have this male camaraderie—I talk like you do, I'm one of the boys, this is why you elected me. He brings himself to what he *thinks* is the level of the group, never thinking that maybe he can bring the level of the group up. Guys come in to work and say, "Oh geez, you should have heard the joke I heard last night at stewards class." So how can I say to women, Go to the stewards class, right?

Donna Levitt was never interested in becoming a foreman, but from the beginning of her apprenticeship she took on an active role in the union. She took photos for her local's newsletter and became involved with a group of members that ran candidates on a progressive platform. She picketed against the two-gate system that allowed nonunion workers onto "union jobsites"; her role as a Bay area labor leader grew and evolved.

W E HAD A RANK-AND-FILE building trades group fighting the two-gate system. It was a fairly dissident thing, because some of the more conservative union leadership was saying there's nothing we can do about a two-gate construction site. It's a union-busting technique in construction where they set up a neutral gate, and the union can no longer effectively picket the jobsite, they can only picket one entrance.

What we would do is—as a group of rank and filers unaffiliated with the local—we would picket, *as concerned citizens,* on the gate where the union couldn't legally picket. A lot of the unions were scared to death with having anything to do with us, because eventually somebody was going to be sued, that we were acting as agents of the union and doing what they legally couldn't do, which eventually happened. It was kind of outlaw behavior, but we thought it was necessary to save the labor movement. Some of the union leadership agreed.

I never imagined I would ever work for the union. I had been elected to be the delegate to several state conventions—but never with the support of the leadership of the local. I remember at one point before I was even journeylevel, our women's group was trying to get something passed at the state convention to allow an official women's committee in the union. Even though I couldn't be an official delegate, I was sent to lobby for this—which we never got. I learned a lot about the politics of the union. I think the next convention I was elected.

When the Carpenters International mandated the merger of the three San Francisco locals into Local 22, the union also did away with elected business agents; the district level made the appointments to local leadership positions. After nine years as a rank-and-file activist carpenter, Donna moved into an official position with the union, the first woman in the district to be hired as an organizer, and later, business agent.

THE MEMBERS WERE TOLD BY the leadership of our district council that even though business agents wouldn't be elected, we would be hiring rank-and-file activists and women and minorities, and that this was going to be a good thing. I think we have a fairly enlightened leadership at our district level. They decided to give me a shot working for the district council as an organizer.

I got to understand who was hostile and who to stay away from, but it was nothing new to me—just kind of like being on a jobsite. After a couple of years, my title was business agent, and the hostility that I faced from two supervisors at my local was brutal. They wanted me gone. I believe they set me up with that intention. I was written up several times.

Once at a union meeting when a new president was told how to railroad the meeting and adjourn the meeting before it got to any open discussion, I lost my temper and started yelling at him. I was written up for using profanity at the union meeting—like men don't do that all the time! I was involved with a grievance for a Latino member, who was not being paid properly by a very crooked contractor. I believe that the business agents allied themselves with the contractor to set me up. Anyway, to make a long story short, the two business reps that I'm speaking of, in the two years that followed, were each fired.

When the senior business rep was fired, Donna had to decide whether or not to assert herself to become his replacement.

THEY WERE GOING TO MOVE in the senior business representative from our industrial local to be the head business rep at both locals. I went to the head of the district council and said that I didn't think that would be appropriate, that I could handle the job. He was worried I didn't have the experience, and I'm sure being a woman didn't help. We probably have the greatest degree of unionized construction in San Francisco as any city in the United States—it was big shoes to fill. When I first said that I didn't think it was appropriate to bring an outsider in there to do this job, he said, "Well, if someone's interested in this job, you better step up to the plate." I thought and thought and thought, and I knew that I could do the job and I knew that I was the best person to do the job and I finally got up my courage to go and tell him that. And he knew it, too. I didn't have to sell him on the idea at all. He just said, "Pull up your socks and let's go."

One thing that I'm proudest of about being a part of the new leadership at the union is that probably our biggest change has been to run a hiring hall where we try to distribute the jobs as fairly as possible. Every dispatch that goes out of here is posted. All the books are open in regards to dispatching. To the extent that our contract allows it, we demand that contractors go through the hiring hall, which is also new. It's partly because the economy has gotten so bad in the last few years that we have had to demand that, or too many people would be out of work, period. So we're forcing the contractors to use the hiring hall to try to share what little work there is as fairly as possible.

Union meetings are much more constructive. Every month I give a very thorough report about what we've been doing, what our problems are, what our challenges are, what our successes have been, and the whole tone of the union meetings have changed.

There are three business agents at this local now, plus organizers who work here, and all of us are committed to diversity in our union. We think it makes us stronger, speaking up on the jobsite, in support of women's rights and people of color, should gender or race issues come up on the job. One of the business agents here, I'll see his reports of what he did during the week and I'll notice in there, "Stopped to check in on Patricia to see how she's doing." Patricia is a brand-new apprentice on her first job. Those kind of things happen naturally around here now. I hope that it makes a difference. It's not just me. It's not even mostly me that does that.

One thing that we do is that we have a good relationship with the agencies that enforce affirmative hiring, the Office of Federal Contract Compliance, the Human Rights Commission—that's the city agency for redevelopment projects. When contractors are not in compliance with meeting those goals, we often alert them. We work with them to make sure those goals are met. Contractors are allowed to request people by name, up to 25 percent of the crew on any jobsite. We let contractors know to reserve those requests to meet affirmative action goals.

We encourage stewards to come to union meetings and, hopefully, that message comes across at the union meetings, that we appreciate the diversity among us and that we're here to support each other. There is an orientation for apprentices when they first come in the union, which I lead along with the apprenticeship coordinator. Hopefully, it means something just to see that it's a woman that's the head business agent at the union. We stress their unionism and what it's about as far as supporting each other and working together towards our common good. I know that there is a policy that's read also at the orientation, from the coordinator, about racial and sexual harassment is not allowed at the school. Our union constitution was just amended to include racial and sexual harassment as an offense that one member can bring charges against another member for. The San Francisco carpenter apprentice program is 56 percent minorities and 11 percent women.

Most days in my job now, I feel like I've been able to do something that's made the union stronger or made a difference in somebody's work life, and that's as satisfying as it was to be able to look at the buildings that were built. After five years working for the union, I still fly out of here with excitement to go set up a picket line, and I hope that that kind of excitement stays with me.

When I first got this job, there was some resentment about me personally, and I had to weather it. I'm definitely in the first wave here. My partner at home, my biggest support, is also a member of the union and knows the players and knows my history. He was a business agent in the Carpenters union and now works for the drywall trust fund as a field representative. Really, I wouldn't have made it through the last four years without him.

But I do get support for being in the position I'm in. Sometimes I hear it— that women are pleased and thrilled and surprised when they come here looking for work, and particularly when they come from another local from some other part of the country.

Nancy Mason entered the Seattle local of the IBEW in 1979, along with eight other women. The affirmative action push from the county and state that began in the mid-eighties supported her and other women electricians into positions of leadership that would have been unimaginable in most other trades or regions of the country. In 1990, Diane Maurer was working as a general foreman; Melanie Sako was chair of the Joint Apprenticeship and Training Committee; Jennifer Balliet was a business agent; and Nancy was the training director of the apprenticeship program.

I THINK WOMEN ACTUALLY HAVE FARED a little bit better being named to supervisory positions than men of color. From what I can see. When I went to the Convention Center in 1988, I was the first electrician, so I became the lead electrician, ultimately became the plant facility manager, and then, in '89, became the training director. I was only at the Convention Center a year and went three steps. But they were actively pursuing, on the state level, an affirmative action recruitment plan. I was first hired and first promoted.

As training director, Nancy put effort into the recruitment of women and minorities, equal access to training, and a safe and respectful work environment for apprentices. The state set a standard of 22 percent female for apprenticeship classes in all trades; the apprenticeship class indentured by the electricians local in 1991 was 27 percent female. This was accomplished by changing procedures, active outreach, and a unified effort from many members of the local.

WE TAKE APPLICATIONS EVERY TUESDAY and Thursday, and I think that if you want to have a broad spectrum of applicants you've got to take applications all year long. I think that has been a much more inclusion type of a practice than a two-week window at some obscure date in September or October. If you happen to find out the week after those dates are gone, your next opportunity is two years later and no, you're not going to wait around. If a program wants to increase their numbers of women, they need to accommodate the application process.

It's pretty amazing for me to go out as a woman journeyman wireman who's also the training director and say, "Hey, you can do this. Women have done this. Come on down."

I think things are better, that we have a lot more members who are truly giving a better shake to training women and people of color in the apprenticeship program than fourteen years ago when I was an apprentice. That's something that I currently am real tuned in to as I go cruising around jobsites as the training director. If anyone calls—and that is not even specific to women and people of color—but if someone calls and says they're always cleaning out the dry shack and they put in two-by-four-foot light fixtures for four months, and especially if it's a third- or fourth-year and they say, But there's a first- or second-year working on the motor control centers, I actually make a call to the contractor. And that would have never happened when I was an apprentice. Sometimes I ask the company to change that, and sometimes I go back to the apprentice and go, "Hey, everybody's had their turn and this is what's available."

One of the things that I *can't* control is the day-to-day on the job stuff. And that really is going to have to come from the field supervision, the actual foreman in the field. In the foreman's training—that's a requirement in our contract for all foremen to take—they have a night of sexual harassment/racial discrimination which I now teach. It's basically a "here's how we're all going to get along and here's the laws and here's what I expect you to do. And, if you don't, these are the consequences. But I'd like you to think about it as a union of hearts and minds, and that you want these people to all succeed—and become good union members and pay your pension." The advantages are that a field supervisor with eight, ten folks in their crew can nip some of the conversations, the innuendoes, get the posters down, take care of some of the lunchroom conversations and the off-color jokes right then and there so that it doesn't escalate into a big, full-scale racial or sexual harassment situation.

I think just the sheer number of women and people of color that are in this local has really helped to be a positive force for women. They have some role models to look at. I think the most empowering thing we can do for women entering the trades is to have women who have made it be those role models, and in fact be mentors if you can possibly get them to spend the kind of one-on-one time with a newly indentured apprentice. We have now in fact a young woman who graduated two years ago who has gotten the E Board to provide childcare at the general meetings, which is an astounding thing. But that was something that she never got, and she turned out and worked to get that.

The achievements of Local 46 IBEW not only provided a model for other trades in the area, it created the opportunity for women in leadership positions in different trades to work cooperatively. Ironworker Randy Loomans had worked as a foreman and was one of seventeen members of the Washington State Labor Council when she became steward on a huge Boeing job in the Seattle area.

IT'S A BOEING BUILDING THAT'S a $120 million job. It's got computer floors, it's generated by big power labs, so there's been more electricians than anybody on this job. I was a steward and we wildcat struck the company.

The six of us stewards met every week to discuss problems we wanted solved on the job. They were safety issues. We basically called an all-craft safety meeting at 8 o'clock in the morning.

Five hundred and fifty employees showed up. I didn't get up on the crane and speak because the three before me said everything that was needed to be said. Then the [general contractor] superintendent got up on the crane to try to talk, and it was like a screaming match. Because everybody was angry. He got down and asked if all of us stewards would come into a conference with him. Well, we went into this conference. We told him, first of all, we'd just like you to get a road into this job. All we asked for was a six-foot walk path, because people fell and hurt themselves walking. So in this meeting they decided, okay, we will give you a walk path and we'll check out all these other safety problems. We thought, Wow, a victory! We've won! Well, the three stewards that got up on the crane—one safety man and two stewards—got fired the next morning. So, we just—none of us went to work.

I explained to my ironworkers what happened, and they go, We're with you. I says, We were all six involved in this. And I went right to [the general contractor] and said, "Why are you guys doing this? Because you don't like what we say you can just fire our stewards and our safety man? Why didn't you fire me?" "Well, you didn't get up and speak." I said, "Because they were speaking for me. They said everything we had to say." He goes, "Do you know what that little safety meeting cost? It cost $25,000." I says, "Well, then, you tell me how much it's costing to have all these men out for eight hours, 500 men. You need to get that steward hired back."

The very first day one of the stewards was hired back, the one from the mechanical. But they still wouldn't hire the safety man and the steward for the electricians. They tried to conquer and divide us all, saying they were going to lay us all off. For three days this went on.

It was quite a monumental thing for me, because the men stayed behind me. 'Cause the [union] hall even came down and had to say that we had to man the job. And they even [said], We're going to appoint another steward. Well, nobody'd take the stewardship.

Anyway, to make a long story short, they hired back the two men. It was like a victory for labor, I thought. Don't dare think that you can fire our people because you don't like what we have to say. It got in the paper and everything. I felt really supported by the men on that one—you're hitting their pocketbook.

It was a stand that I was willing to lose my job for, and didn't blink an eye about it because I felt so strong about it. Jennifer [Balliet] asked me to come down and speak to the electricians at her hall to tell them what was actually going on. She just closed the door and I told them everything. The way they manned the job was that they would send people out per request and then they'd, "Well, where's the steward?" "Oh, we don't have one." "Can't work without a steward," and back to the hall they'd go. Then Friday night it was settled.

To top it off, the guy that we fought to get back was thinking he wasn't going to come back to work. He wasn't there Monday and I called down to Jennifer and I said, "Where is he? These people stayed out of their job so that this man could get his job back." Jennifer goes, "Well, we're trying to talk him into coming back. Would you talk to him?" Because this is a man that I've met with every week, so I felt real personal to him, too. He got on the phone with me and I says, "Everybody on this job's asking where you're at. These men all feel like they've stayed out for you, to get your job back. It'll be all for nil and void if you don't come back." So he goes, "Well, you're pretty persuasive, Randy. Okay. Tell them I'll be back tomorrow." It ended good, too, and I felt like I had a part in the ending of it.

Expansion Joints

> When we had our pre-apprenticeship training classes I used to tell
> all the women getting into it, I said, If you are not an assertive,
> aggressive individual, if you are not that type of person—you will
> learn to be, before you're through with this. To survive, you have to
> pick up on those skills.
>
> —*Kathy Walsh, Kansas City*

Winter '93–'94. The economy's tight. Those who get laid off face about a two-year wait for another union job. There's pressure to complete projects with fewer labor hours and less concern for craft.

I'm working for a woman-owned company. But that doesn't stop one of the foremen from informing me that he has personally worked with all of the women electricians in our local (more than a hundred!). Only one, he tells me, is competent at the trade.

At coffee break I find myself listening to vulgarities I hadn't heard for a dozen years. Jokes about pedophiles and homosexuals, jokes that the one black on the crew must be a drug dealer, jokes about the size of girlfriends' underwear. Mostly initiated by first-year apprentices. One of the foremen is eating a pumpernickel bagel with cream cheese. An apprentice asks, "How does it feel to have something hard and round and black in your mouth?" Loud laughter. Someone from the office is in the shack, he starts talking to me about the weather.

I'm remembering my interview to become an apprentice in 1978, being asked about my ability to work with men. I'm wondering whether these apprentices were questioned about their ability to work with women, or people of color. A lot's going on in my personal life. I'm wishing I could just come in and do my job, that either the union or the contractor had made the workplace ready for me, so I wouldn't have to choose between the same too-familiar choices—silence or speaking up—and their consequences.

With all these years in the trade, women still seem to have barely moved from the starting gate.

—Susan

Tradeswomen's perceptions of the industry were not static. The expectations each woman brought to the beginning of her apprenticeship were met—or not. Shifts that came about in the nature or organization of the work were positive for them—or not.

Temporary accommodations that women made as pioneers breaking into an inhospitable workplace had to be reconsidered if the adjustment required turned out to be permanent. And tradeswomen themselves changed—their situations, priorities, tolerance levels. The work itself brought on changes in women—desired and not.

Decisions of journeylevel mechanics to stay in or leave the industry—even when arrived at in a moment—were based on each woman's complex weighing of positives and negatives that included her evaluation of whether or not the various ways the industry was calcified or flexible suited her long-term needs. The transient nature of the work—shifting from one jobsite to another, one contractor to another—created many points of reflection, when one could envision the next worksite with hope or dread.

As she reached her forties, electrician Sara Driscoll found the many silences and invisibilities she had accepted for years less tolerable.

PART OF MY PROBLEM IS that I felt isolated as a lesbian. I felt like, I can't even come out and be who I am, I can't make friends here. I can't identify myself out and feel safe. The fact that I was Irish and I was kind of tall, those were somewhat protections. But my lesbianism was not a protection, and it was something that was really scary to me, because even though I got to work with a couple of people very, very closely, I still never, never told anyone that I was a lesbian. I never said those words. I never came out to anybody the whole time, the whole ten years I worked, and it felt very, very isolating. Even when I wanted to come out, I had this instinct that said, this guy is going to get drunk someday, he's going to be in a group with the guys, and he's going to say, "You know about Sara?" And the next thing you know on the job, I'm going to have a ladder kicked out from under me or—I don't know. It just didn't feel safe.

I spent the first three years talking about how for graduation, I was going to bring my friend Steve, which is who I brought. It felt really silly. Even though it was fun, we had a great time, I on some level actually regret that I did that. And I resent that I had to do it. Or that I felt like I had to do it. I regret that I didn't just either go by myself or bring a woman.

I ended up leaving the summer my mother died, and I think that also had a lot to do with it. My mother died in April of '88. And I left in August of '88. I spent that summer on a job that was really nearby and I could ride my bike to the job. I'd go to eat lunch in the Arboretum every day, but I was by myself.

Some of the other tradesmen would go there, but I never could really have conversations with these guys.

They talked about things that I really didn't care about. Here I am 40 by this time, it's almost ten years later, and I really didn't have any tolerance for having to be around things that I really didn't give a shit about. You know, sports—and it's not like I don't even care about sports, but I don't care to the degree they do. And I don't gamble it. And I don't care about the lottery, which is a lot what they talk about—numbers. I would have a hard time scoring somebody on the job that I could really schmooze with.

I just remember having a very hard summer, having lost my mom—which was a really hard thing for me—and feeling totally no space on that job to grieve and be there and just even have anybody recognize that perhaps this is what happens. A couple of times I would just break down in tears and had to leave the job. I didn't say, My mom died and I really want to talk to you about it. I never felt that kind of space, or comfort, or safety.

I got transferred from that job—this is how I left. As I was getting transferred I said to my foreman (I know they have a million jobs going on downtown), Can you please see if I can't get to a downtown job so I can ride my bike in and I don't have to travel. I live in town, I'd like to stay in town.

No. I end up getting sent out of the city and I was like—come on! This isn't really necessary. So I go to this job—and of course the foreman is some cadet who doesn't know what he's doing, there's no stock, the work that had been done previously was wrong. We had to redo all this stuff from other people's work. It was bad enough to have to redo your own, but to redo somebody else's is really an irritant. I was, apparently, above my eyeballs with—enough of this!

I was sitting there having coffee in the summer afternoon and some guy from one of the trades comes over. Comes over and he puts his hands on his knees and he says, "So, how are you liking it in the trade?" I said, "After ten years, Jack, it's just fine." "Oh, I thought you were a helper." I said, "Do I look like a helper? I'm 41 years old. Give me a goddamn break. Why assume that?"

I had absolutely no patience. You know, in the old days you'd say, Oh, no, I'm not a helper, I'm a journeyman now. You'd be really nice. It's like—no! I really don't have the time and energy for this anymore. So it was at lunch that day that I said to this guy I'd been working with, I'm outta here today. He said, Oh, sure you are, there's no work out there.

And I did. I put my tools on my shoulder that afternoon. I walked out the door and I never looked back. I called the shop in the morning and I said, I'm outta here, send me my check.

My body is really happy I'm not doing the work. My spirit is really happy I'm not doing the work. Your lungs get attacked constantly. Your muscles are toned from the work, but they're not flexible. I used to do yoga every day after work just because I'd come home and feel so bound up. I'd have to stretch out and free up. So I don't miss that part of the work. I miss working with the women. And I miss sometimes the production of the work, actually doing it.

I've watched the process change and that feels good. I've also watched it remain the same, which is a little depressing—the status of women in the union, women in the trades. I feel like it's changed in that it's not so weird to have women be doing this work, so that on some level we can never totally go back. Even though Rosie the Riveter happened in the forties, women did get shoved back into the houses. Somehow I feel like this time that's not going to happen, even though there's no work out there or there's no encouragement, the actual understanding of women's skills and ability I don't think can ever go back. I feel like that's been an important thing that's happened.

Where it has remained the same is in the recruitment policy. It is still highly resistant to women, even though they know we're smart, we're good, we're reliable. Everything. That whole we're-taking-jobs-from-men mentality hangs in extremely tight. Especially when a recessionary time comes in. I think they've changed individually within the industry. I think a lot of hard-core guys in the trades have changed their perceptions and deep down in their hearts know we are right. Know that this whole thing is bullshit about women belong somewhere or not. I've had guys say, I'd hire a woman over a guy any day, now that I know the kind of productivity I can get, the quality of that pro-ductivity. But there's not a lot of men out there banding together saying, We support these women. That's just not their culture. And that's a shame because they're losing out, and the women are losing out even more.

I don't want to *not* be an out lesbian anymore. I don't do this closet act very well. And I did it for a very long time. It got to be too long. Before I went into the trades I was an out lesbian in my community, in my work, and everything. And to have to closet-up went against my grain—a lot—and I held onto that for ten years.

I really liked what I learned, and I liked what I experienced a lot in many, many ways. I'm really glad I did it. I still have the skill. I still can use it. It's not like it's gone to me. I keep my license alive. I won't ever let that go as far as I can tell right now.

In Cleveland, Cheryl Camp found the varying nature of work a stimulus for her own growth and flexibility as a mechanic.

I DIDN'T HAVE THE CONFIDENCE I think that a lot of the guys had when I finished the apprenticeship program. I didn't feel like I was very fast or really efficient enough in my job. I could do the basics, but I saw guys that were in my class come out, and it just seemed like they were so much more experienced than I was. And I couldn't understand why. I felt maybe it was because of my gender. But then I found out that most of the guys had done this type of work in some way or another through their uncle or their brother, 'cause there's a lot of nepotism.

But, after you've done this work for a while, I'd say if you've been in eight or nine years, then you start to develop that confidence in yourself. Just lay the job out, tell me what you want me to do, and I can do it, I'll figure out something. It's all in how you perceive a job and how you yourself as an individual are going to gain your confidence in yourself and in your abilities. 'Cause you can have the ability but lack the confidence—you don't have a good combination there at all. But, you know, I have confidence now.

What I have found is that I have gotten to meet a lot of people, which has helped me as an individual to grow. Because you interact with so many different people, so many different personalities, and that's been beneficial for me. And working on different work sites, different jobs. I've done so many different types of things in the electrical field—it's not just the basic bending pipe and pulling a wire in and putting on devices. I've worked in a mill, I've worked in hospitals. About the only thing I haven't done is brand-new homes. My apprenticeship and the work that I've done has been pretty diversified.

It's the time off work, though, that's hard. When work gets low, and I'm laid off, 'cause I don't like staying at home. I can't sit at home. But I can't travel, either, like some of the guys do, go out of town and work.

I wouldn't advise it for everyone. My sister, for example. I have a sister that can't stand the thought of dirt under her fingernails or her hair getting messed up or dust in it or something. And if you're a female like that, then you're not going to make it in the trades. Because you're going to get dirty, you're going to get dust all over you, you're going to be placed in some situations where you wonder, "Why am I here? I don't believe I'm doing this." 'Cause some of the jobs that I've worked on, I tell you, I've likened it to Army boot camp.

I was running a pipe run above a finished ceiling and it was plaster inlay, and we had the metal wire mesh. They just cut a hole in it, an access passage. We had to go up in there, my partner and I. They picked us 'cause we were the smallest ones. We went up there and we were crawling around in between and underneath duct work. In some places it was so narrow that you had to lay down on your stomach and scoot underneath this whatever-it-was. There was steam lines. I just laughed.

I said, "I don't believe that we're up here doing this. I used to be a secretary, I had to wear dresses and stockings and heels, and look at me dragging a trouble light." 'Cause it was dark. You had no idea what else was up there with you, you know. You got your light and you're dragging it and looking, shining the light around. Then you'd crawl a little bit more and then shine the light around. It was wild.

You never know what you're going to do at any given time. I think that's part of the adventure of it for me. You're not with the same people all the time, you're always working someplace different, and it's different types of work styles. 'Cause every company has a different method that they want you to use to do their work. It's really, really a great advantage to me. But it's not for everyone. Some people like to have that security of knowing, well, this is my desk and my chair, and these are my plants and I know I water them every Thursday—and they could not see themselves doing what I do.

I don't see myself 45 years old crawling around places, but I do see the potential for other aspects. Becoming a contractor, or an estimator, or an inspector. That's what's great about our field, too. 'Cause there's other avenues open, you don't always have to do the physical end of it. Once you learn the technical end, then you can go into that realm. Eventually what I'd like to do is own my own company. Probably small-scale, though. You have to have a lot of money to start, so that's what I'm working towards.

For Angela Summer, a plumber in Seattle, the work did not hold up as well.

IN PLUMBING, YOU CAN PUT in all these big pipes and stand back, and every day you could see this concrete thing you had done all day. You had put in this many feet of pipe, and you could see this long run, so it was real job-gratifying. I was kind of like young guys with cars or something. For a while there I really got off on working with this big stuff and drilling holes and doing all this.

I feel like I've just gone beyond that now. I've come to realize that it's more of a glorified labor and grunt job. Once you've mastered something or learn to do it, you do it better and better every time, but it just starts getting old. It's like, okay, I'll run copper pipe, I'll run this kind of pipe, I'll run gas pipe. It will be interesting for a while. I'll be doing something I haven't done for a number of years, but then it will be the same old thing again.

In construction, you're out there all through the winter, a lot of times in a bare bones building, just the iron's up or something. You're drilling holes out there and putting in pipe, and the wind's blowing, and you're up in the twenty-something floor and freezing your butt off. And if you're lucky enough to be doing copper work, you can torch your feet every once in a while and keep yourself warm. But my large toes of both feet are kind of numb from doing construction work, being in the cold so much. I have pretty bad circulation.

In Boston, Deb Williams found changes in the industry a welcome challenge. Rather than craft values decreasing over time, she experienced a return to more labor-intensive techniques that had been little used for decades.

A T THAT TIME MOST OF the things they were doing was pretty cut and dry. It was all sheetrock, and then we went in to spray paint. Ninety percent of it was masking, and then the spray painter would come behind you. But now, I'd say probably '89 till now, they're bringing all the old things back. Now everybody wants all these meticulous things, especially wall coverings, marbleizing, graining. They're doing the special spray paintings for the poly mixes—a whole different ball game. They're going back to where it was in the thirties and the twenties or like the fifties.

The men I work with are in their fifties. They all mix colors, mix and match, they can take any color and mix it, some better than others. They can do wood grain and marbleizing. Years ago, when you were a painter you taped, you painted, you marbleized, you leaded. These guys would paint it with Dutchies. They were twelve-inch paint brushes—never even heard of a roller. They didn't even make rollers legal till 1978—well '77, '76 there was like a seven-inch roller. Now it's unlimited use of tools. They used to limit what kind of tools you could use. Back then these painters learned everything. They used to mix the lead with their hands in the paint. You hear the old-timers talking, I'm like—wouldn't I love to know all that.

The union now is doing what they call journeyman upgrading, because a lot of the journeymen, say, my age and maybe a little older, and maybe just a little younger, never got all this special stuff. I mean 99 percent of my work was always brush and roll. Then I started getting into Symphony Hall, and the old-timer that was there, he was supposed to teach me some of these wood grains and things. But he used to just find other things for me to do. They would give you a week of it or a couple of weeks of it in apprenticeship school, but they never actually developed it. Now I'm picking all that up. I applied myself more in journeyman upgrading than I ever did then.

Sometimes even a small change—because it represented a shift in trust—made a big difference. Only as a journeywoman did Lorraine Bertosa change her choice of work clothes.

T HIS ONE GUY USED TO model on the side. Tripped me out. He'd model for Higbee's and local stores around here. The best-looking construction worker I've ever seen in my life. He had a beard and it was always impeccable. He had a little scarf he'd always wear. I started watching him. Up to that point I always thought you had to wear these clothes 'cause they get dirty. You gotta wear the dark plaid, you gotta wear the gray, the black—I started buying pink sweatshirts and yellow sweatshirts. It was like—you go home, you throw it in the wash. So it's not perfectly clean. Buddy down the line after that bought me a pink hardhat. It was like, Screw this, you don't have to wear this drab color out there. It was a lot of fun. You could look nice going to work. So at the end of the day you're all greasy—what the hell! Who cares. Tide cleans it.

Becoming a skilled tradeswoman required enormous personal change.

I THINK WHAT I FOUND MORE than anything was, I wanted to be all this involved, but here I was a single parent with all this emotional turmoil going on inside that I wasn't even in touch with at the time. There wasn't a place for *that* in that arena. Nobody was talking about personal problems. I kept wanting to take care of my personal problems by doing more and more activity outside myself. I wanted to become union involved. But I didn't realize I had to stop, and start taking care of Lorraine inside. It took me a number of years to realize that. I see a lot of young women today who don't have that self-doubt I had. They feel confident about who they are. It's not across the board, but the younger generation has got some of that, that I don't have. That I still fight to have.

It was real positive for me because it required so much, that I had to winnow out these other focuses in my life. I had to stop focusing on family members who were still drinking at the time. I had to start making decisions to *not* focus on them and not worry about them and to let them take care of that themselves. When stuff happened in my relationship, I had to still keep making that choice—if I wanted to stay in the trade, and I did. Everything else had to move out of range. It took a lot to do this.

I'm watching my friend go through residency right now, this doctor degree. Everything else has to take second fiddle to that. I had to do that, too, but I didn't realize that's what I was going to need to do. When I teach now I try to express that, because there's no need to go in and think you can just do this as a sideline and pick it up. The men have had practice, most of them, with their dads or whatever. Maybe the woman has gone through shop classes by now, or hopefully there's some exposure. They've been playing with something, with wood or electrical or whatever, so that it wouldn't be as Greek as it was to me—which was that added burden.

It enabled me to do something I had never done before—to focus in on me. And it took all those years to do that, because I was a hard case. I was.

From the security of her job as a civil service carpenter for the City of New York, Irene Soloway could look back on her apprentice years with a changed perspective and analysis.

I WAS APPROACHING IT LIKE A good feminist—like what I thought a good feminist was, probably not like a good feminist. But I considered them the enemy, you know. It was all about proving something, and I was very, in retrospect, very frightened of the situation. But I never admitted that to myself. I always thought more that, I'm going to get this job, I'm going to survive this. I'm going to show them. It took me a while to conceive of even having friends on the job.

I felt like I was a creature with two heads, that I was like an alien being and they were alien to me. But instead of just seeing it humanistically as, yes, I am alien to them, they are alien to me, I thought politically that, you know, they are the ones who were trying to keep me out. There wasn't very much middle ground in my mind, but it took me a while to figure out.

The fear was just the fear of being an outsider, danger on the job, not knowing the appropriate ways to handle situations, since they were totally new to me. Like when the work itself was physically very difficult, not knowing what limits were reasonable—and I think a lot of women got exploited that way. I just didn't know what standards to set for myself, so it was frightening. And it was a culture shock for me, since I came out of a gentler environment. I had to be very angry and very self-righteous—that's what I thought.

It's so nice to look back and to see how far I've come in terms of now, I can set my own standards of behavior and I can insist on them. And I realize I have a lot more power in a job situation. Particularly as somebody who has an education, I have more power than a lot of the men. But at the time, I thought I had to be some other person to survive on the job, so that was very uncomfortable. You know, do you laugh at their jokes or do you get angry—instead of making your own jokes and saying, Fuck you—that type of thing.

The first job I was on, they asked me to sweep the shanty out and keep the time, keep the records. It was their approach—they had to have a woman on the job—and I thought that was pretty outrageous. And I said, What, are you crazy? If I was going to do that, I would come to work in an angora sweater and sweater clips. I mean, why did I buy these work shoes for?

So, like that, you know—foremen who don't know what to do with you. Or just being given shit work and coffee and that type of thing. Of course, now I understand that that's sort of normal. But at the time I was offended. I felt so humiliated of that whole procedure. I wish I was a little wiser, I would have been more graceful about it.

I don't think the men that I worked with were so terrible, I didn't have any traumatically terrible experiences. It was more the feeling like I wasn't being given opportunities to learn and that, when you walk in the shanty everybody gets real uncomfortable. That type of thing. I never felt like I really got the good jobs and got somebody to work with that could teach me. Maybe my third year of apprenticeship, I started to get some better jobs. Really a lot of what I went through was just normal, but being so anxious about it, I always felt like I had to play catch-up.

This is what I've learned. If you were a woman who kisses ass, you will be in the same position as a man who kisses ass, you'll get work. The submission and playing the game is more important than gender in the carpenters union. That certainly was my experience, and I've seen it in other locals.

That's one of the reasons why—if you want to talk about women's groups and why we haven't gotten very far in New York—I think the men don't have it real good. That dominance and submission theme and the lack of democracy affects the men greatly. I think that it's great that women can set a higher standard, but I think it's going to take a lot more.

The issues, then, became a lot more for me issues like where I saw how the structure of the union wasn't working for the men. Submit or be left out in the cold, that type of thing.

Mary Michels recognized that her decision to have a career as an ironworker required personal changes.

I WASN'T USED TO THE WAY they joked around and talked. A lot of it could be very offensive, in their remarks that they make. I'll never forget this one time I was crawling across the beam—because I wasn't about to walk on it. It was really thin. To this day, I would crawl on it—they call it "cooning it." Then these two guys—it was the way they said, "Hey, what would it take, a couple of drinks? A fine beef dinner?"

I was very offended by that, but they just laughed it off. They thought it was the funniest thing in the world to say that to me, especially when I was in such a vulnerable position, crawling across there. I was mad for a few days and they just, you know, didn't think it was a big thing. Then I kind of saw, I am not going to let these people get to me.

I'm very calloused against those kind of remarks, I don't let them in, I usually don't hear a lot of the stuff they say. I keep it out because I like the work. I'm not going to get offended by it because if I get offended by it, I'm going to end up quitting. So I have made myself calloused, and I know a few other women are the same way, we just ignore it.

For Melinda Hernandez in New York City, pioneering carried heavy costs.

YOU GO ON A JOB and, the men are immediately testing you. And so another job is another turmoil, another stressful encounter. How much can you go through, before you snap or you lose it. You know?

You become someone that you're not. Basically I used to be a very nice, easy-going person, very friendly, very open, very trusting. And I've become very hard and very distrusting. I'm not that innocent anymore. I lost a lot on the job. I became a lot like them. Anybody mess with me I was ready to just jump. My family was afraid to have an argument with me. Even if I was wrong they wouldn't want to say nothing, because oh, God, I would blow up like a time bomb.

It was constant pressure, constant stress, constant being picked on, constant being set up—it's too much for one person. About four or five years ago I met a woman, she was a traveler, and she said to me, If you had to do it over again, would you do it? I said to her, Yeah, sure I would. And she goes, Not me. Not only wouldn't I do it, I would discourage any other woman from doing it. It takes too much of who you are. And at the time I felt really bad for her that she felt that way. But now I know exactly how she feels.

I wouldn't encourage no one to go into the trades at this point, because it's a damned if you do, damned if you don't situation. If there were *more* women then I *would* encourage them to go in. But in order for there to be more women you've got to have women come in. You get paid well, but you pay the price. It's a high price to pay. You don't really get acknowledgment or credit or gratitude, you don't get these nice things, these human emotions that keep you going. It's a very hard thing. I think I deserve more. Not just more in my paycheck. I deserve more in life.

I used to say, I love my job, and I meant it 100 percent. I loved learning it, I loved doing it, I loved bending pipe, pulling wires, I was into it like if I was going to the playground in the morning. I really was having a great time learning the trade. But they took the heart out of me.

The craft itself, yeah, it was wonderful. There is a fair share of grunt work, but I think our trade is the most you use your brains in. It was always a thrill to me to be at a new jobsite and doing new work, whether it's a hospital or a school or a two-family house, or a subway tunnel—everything to me was exciting, it was an adventure. I like my work now, but it's not like I used to feel. I used to have a real love for it.

I lost heart in the work because of the harassment, all of these emotional things that you need—feedback—you don't get. The disrespect. The inconsideration.

The one change there has been is that there's no pornography on the jobs. It's unacceptable anymore. When I came into the trade, it was wall-to-wall pornography. I remember I was on a job, I was there for a week, and I was getting a coffee order for thirty-five men, I was the only apprentice on the job. I was getting their lunch order too, and I was getting their afternoon coffee order. So how much time do you think I was learning, right? I didn't have my own shanty either.

The foreman asked me to clean up the men's locker. Now, I'd been sharing the locker with them all week, and it's wall-to-wall pornography—I mean, really gross. I said, Yeah, sure, I'll clean the locker. So that day, after I got coffee back to the men in the morning, I went downstairs and I commenced to rip up every single photograph and picture on the wall, every article and everything, and I threw it all out, and I ripped it up so it couldn't be put up again. Well, that afternoon I came back from lunch, and boy, there was like a crowd, like a stampede in the corner, just waiting to get me and lynch me. How dare you do this? Who do you think you are? She did this, blah, blah, blah. And the foreman's like, Whoa, take it easy guys.

I think it was second year. And the foreman said, Melinda, why did you do this? I said, You asked me to clean up the shanty, so I cleaned up all the filth in the shanty. And the guy's like, You come into this industry, it's a man's industry, and this is the way it is, and this is the way it's going to be. And I said, No, times they are a-changing, honey. Those pictures go up and every single one of you goes out for your own coffee. They wouldn't even talk to me. The super of the shop was called in and everything.

Those were the kind of women that we were when we were coming through. A lot of the women that followed are not like that. A lot of the women that followed don't want to make any waves. I always thought it would be different, immediately after the first four graduates. I thought that the treatment of the women coming in would be different, and I thought the treatment of the graduates would be different.

So I'm glad that women are not exposed to that much pornography anymore. I think the sexual harassment might have diminished somewhat because of the lawsuits, when the Human Rights Commission had the hearings—those brought about change. Anita Hill, thank God for her—more men are realizing that there are certain things they should be careful when they say, because they don't know what the repercussions might be. I think that that's changed a lot. It should be completely eliminated, but it's also up to the women to stop it.

Women have to take a stand and say, Look, this is not acceptable. Stop hitting on me. I'm here to do a day's work. I'm not here to go out with you. I'm not here to boost your ego. We should get along as colleagues, but we don't have to get along as men and women, because that's not what this is about. Two people working, that's what we are. Work is work.

In Seattle, where county and state affirmative action standards encouraged the hiring and advancement of women in construction, Randy Loomans saw ironworking improve not only her life, but her daughter's.

FOR ME, AFTER BEING IN the trades ten years now, the men know me. I really feel like I have a large degree of respect from all the men I work with. They may say things about me before they work with me—but after they work with me there's nothing they can really say. It is a shame that you always have to—not necessarily be proving yourself, but you have to be on top of it all the time. Because no matter what they say, men are always going to be looking for that weakness. And that's what they'll site, and then they'll say, "Here comes another damn woman on the job." I have a lot of integrity about that. I work real hard and I usually always outwork my partners. It's been great to see how the men have come around to me, though. I really feel respected. That is really a mega thing.

Randy Loomans and her brother ironworkers stand on newly installed beams at the Green River Gorge Bridge in Black Diamond, Washington.

When I first got in, I didn't even want to tell people what I did. I didn't know what they would think, but it felt like negative to me. So nontraditional. So manly. I think partly because I didn't feel it, either. I didn't feel that I had earned that title, because I had too much to learn. Now, of course, I tell everybody I'm an ironworker, 'cause I feel like an ironworker now.

I try to get my daughter on all the big high-rises I work on. I bring her down and show her what we're doing. She'll come and have lunch with me. She spent a day at work with me before. We just got her in proper shoes and a hard hat and she came. It was a job where I was actually the boss, too. We were setting panels and she watched the whole show all day long. She was just totally impressed. She was scared that I was so close to the edge.

I think it's really been good for her. I think that she sees that it's brought us a good life. I live pretty much how I want to live. I built a new home last year. I think she sees that women's capabilities are limitless, the only people that set limits is yourself.

She's going to college, and I want her to go to college. She goes, "Well, look at you, Mom. You make good money." I says, "But that's the point. I have to go out in the winter, I have to work in the cold to make this good money. I want you to be able to work in an office and make this good money, or work wherever you want to work." She would never do what I do. It's not part of her. But I think she's real proud of me.

Bernadette Gross completed her carpentry apprenticeship in Seattle, where there was a large and varied community of tradeswomen and, eventually, a greater "ability-based" acceptance of women. She returned in the mid-eighties to Washington, D.C., where organized labor and the tradeswomen's movement were weaker.

THEN I COME BACK HERE—you know, it's a time warp. Hardly any women, hardly any support.

I began to weigh out whether I was willing to continue to go through initiation, whether I was willing to have people stand around and apply so much pressure that it really became a choice of whether to stay on a job or leave a job. As my kids got older and my real true purpose for working hard like that wasn't there, the money mattered a whole lot less. I had some marketable skills. I didn't have to take their shit anymore, that's the bottom line. So why would I stay there and continue to be a double minority and to be treated like, Well, you're only here because you're a woman, or you're only here because you're black. And, we know that that means, that you can't do anything and you don't know anything.

But also for Bernadette, her work in the trades had positive repercussions for her children.

ESPECIALLY FOR MY DAUGHTER. SHE'S really assertive, a go-getter, knows that women can do whatever it is that they set out to do. She's in a male-dominated field. She's out in L.A. trying to work from the production side of music and she's seeing a lot of stuff, too. I mean, it exists in the culture, the whole society, and it doesn't matter which area you get into. There's this sort of understanding that it's their territory.

She wrote this really beautiful paper once she graduated from high school that was like a real tear-jerker—you know, seeing me struggle through welfare, through addiction, through all these things. She always talks about how grateful she is that she was given those opportunities, and you know, she realizes what it cost me. Because I would often talk to her about breaking cycles, that what chronic poverty is, is the fact that no one ever gets enough to pass on.

I am a first generation college student and she's a first generation graduate, her daughters or sons go on from there and it breaks the cycle. I told her that you're not going to graduate from college and get a good job and—zappo! That happens over a period of time and you have to be willing to make that investment. I invested in her and she invested in the next generation and then pretty soon someone will, hopefully, sort of edge on out of that poverty range. Because a good job does not break that cycle automatically. There are a whole bunch of things like time management, financial management, all these kinds of things that you never even have a inkling of.

Getting money and using money properly is two entirely different things.

Passionately proud to be a carpenter, Kathy Walsh found the stress wearing over time.

I HAD MAYBE ONE HANDFUL OF guys that I had real trouble with on jobs, over more than a decade of working in the trades. It's easy for me to get along with people. And everybody, for the most part, liked me. I was good entertainment out on the jobsite. I've heard that so many times recently here, it's amazing. "We don't mind having women out on the jobsite because actually they're fun to have around, they lighten things up a lot." But it's like, they always let you know that you were there because they needed a woman on this job. There were only a couple times where I ever felt like I was there just because I was a good carpenter.

I think I snapped.

I had been unemployed, it had to be about '86 or so. I had been called back to work for one of the companies that had called me back several times. They had a carpenter foreman that I had worked with many times who was a real lady's man, hustler. But I had known this guy and how he operated for years. I went out on this job and he was the general carpenter foreman. Also on this job were at least three other guys that I knew very well, had worked with on different jobs and it was like, they accepted me as their peer. They were good carpenters and they liked me being on their crew. That was rare and meant a lot to me.

My general foreman on this job, he was always doing his same old stuff. If he needed anybody in the saw shop working, to where he could be up there hanging out with them and just bullshitting or flirting or not working, it was me. If there was anything that needed to be done where it was him and one other carpenter and could be off somewhere doing something, it was me. I was feeling real leery. I didn't trust him—sexually, physically, man to woman or boss to employee.

So I'm trying to deal with him. And then I got this crew of guys I'm working with, and some of them are really good guys and they can see what's going on, too. They're starting to razz me about it. I'm not in a position—you know, when he says you're in the saw house today, that's what you do. And I didn't know what to do.

I went to work one day and he called me up to the saw house. [I felt] like a child who couldn't do anything about it. And like a whore, because I felt like that was what all the guys thought. He called me up there. And instead of going up there, I went to the gang box and I gathered up all my tools and toolboxes, and walked off the job.

I think about two or three days later one of the guys called and said, "What are you doing?" I said, "I'm not doing anything." "Where are you? Are you working?" "No." "Are you sick?" "No." "Are you coming back to work?" "No." "Why not?" "I don't know, I'm just not." That was it. I just felt like I couldn't do it anymore. I stayed around home and collected unemployment for a while, and then I went to work for myself, just doing little odd jobs—friends and relatives and friends of friends. And from there I went to work for the new Hometown Plan that was approved here.

I'm still proud to be a carpenter. Completing the apprenticeship and proving that I can make it out there as a journeylevel carpenter—no matter what happens I can take care of myself and my kids.

CHAPTER FIFTEEN

Punch List

PIONEERING
for the tradeswomen of '78

She had walked into their party uninvited
wedging a welcome mat in the doorway
for other women she hoped would
follow along soon.
            The loud ones argued
to throw her out immediately. Even her supporters
found her audacity annoying. But once they saw
she mingled with everyone
drank American beer
kept conversations going during awkward silences

and was backed up by law
the controversy
            calmed.

She surprised them.
She was reliable. She always gave her best.
She was invited back.
She became a regular—
        always on the fringe
        expected to help out
                    just a little more.

When she stopped coming
they were confused. Why now? Hadn't she
challenged custom? stared down rumors? ingratiated herself
years ago? so that now her presence was only
mildly discomforting. She never explained.

After all those years hurling back cannonballs
womanizing the barricades firing
only if she saw the whites of their eyes
it was the lonesomeness
                of pioneering
that broke her resistance.

All those silences
> about what mattered
> most in her life
had worn her

like the slow eating away of acid on metal:
the damage only visible over time.

—Susan Eisenberg

— *Irene* —

I think what would change it the most would be enforced federal regulations. I think that's what opened it up, and that's what all the contractors were thinking, Oh this is going to be enforced, we had better do this. All these other issues—if they have to have women, then all these other things will generate around it. So that to me is the main deal.

— *Lorraine* —

The negative piece of the thing still, that I can remember back then and remember now, is the piece of being the only woman. That is a negative piece in the whole damn thing. It's always been such a difference when there's been more than me on a job. It's always such a delight.

Once I worked with a good friend of mine, only once in twelve years.

— *Mary* —

At this point, I would never advise a woman to get into the trades. Chances are pretty slim she's going to make it right now, unless we have stronger laws. Late '80s, I would have advised it, yeah, but now—no.

You need enforcement, period.

— *Kathy* —

It amazes me that contractors, they're always, Well, we'd hire women, but there aren't any. And unions are, We take women but the contractor won't hire them, and We can't find them.

There is no advertising done. There is no PR done. There is no information work-shops, how do you get in, what do you do to get in? When we did our pretraining classes here, just from a couple radio spots on public radio here in Kansas City—

Carpenter Lorraine Bertosa supervises construction for See Jane Build, a joint project of Cleveland's Hard Hatted Women and Habitat for Humanity. Photo by Mark Urbanski.

and we went on a free public access TV show—we never paid a penny for any advertising when we were recruiting for our classes—I still get five to six calls a week from people who heard a thirty-second radio spot six months ago about women getting into construction. The phones would ring off the the hook.

— *Paulette* —

I'm a member of Women Empowering Women. What they ended up doing was having weekend workshops to give women basic hands-on experience with the three major trades, plumbing, electrical, and carpentry. There would be like an all-day workshop on a Saturday. Teach women how to repair faucets or solder pipe or do some basic stuff, usually women who own their own homes and wanted to be independent of repair people.

We remodeled a broken-down house in Berkeley. We renovated it as a house for the homeless and it took months, because we just had to completely gut the place. What we did was use it as a training ground, so that women would come to the site, actually crawl under the house, put the gas piping in, come out and thread the pipe, go in, put it in, measure the next piece, come out and thread it, go in and put it in. Do the sewer line, do the water line, do the sheetrocking and carpentry and paint. I mean, we were on that house for months—the roofing, everything. And a lot of women had rave reviews about it. It's pretty incredible when somebody comes out from under the house and they say, I did it. I actually did it. I mean, they actually put the stuff together. They didn't know they ever could. They never thought about it. Never touched it. It's a great feeling to watch people experience that.

— *Marge* —

I think that all the words about "we'd be open if you really supplied the right people"—overall it's just rhetoric. There needs to be somebody forcing them to take them.

— *Bernadette* —

I'm having people call me and say, Can you get me a trained woman out here? I need somebody with some experience. And I'm saying, You're going to have to take some responsibility to give them some experience.

Everybody's sort of laying into the Catch 22: We'd hire a woman but I just can't find one who has any kind of training. When a guy can walk in off the street with no training, but they assume that he has at least enough muscles to carry him through. And then it's a game of repetition, so after you've done it about a hundred times, you can do it.

— *Cheryl* —

It's not real visible. Like, when I got in, I had no idea that construction was going to be an alternative to college or the military. I just happened to find out about it through a relative. Actually, junior high, I would think, would be where the outreach should be, because they should start preparing in ninth grade for the apprenticeship programs. For electrical, you have to have a strong math background, and electronics courses help, blueprint reading helps, all of those things help build your foundation for you to build on. I feel that we need to touch base with these junior high school students and let them know that construction is an alternative.

— *Gay* —

Even when they do do recruitment, when they do outreach, it is done in such a way that you really would have to be a little crazy to think about doing it.

— *Marge* —

My problem now is being with people who think they're doing women a favor by telling them how hard it is. Yet by telling them that, with not followed by, "and you can do it"—put in the wrong context—they turn people away in droves. They're always saying the trade is so dirty and the hours are so terrible and you're laid off all the time and it breaks our body and all of this stuff. And yet, they've been in it and they stick with it.

— *Nancy* —

One thing that then happens once you get a significant number of women in your apprenticeship or in your local union, like we have now, is that those people grass-roots-wise recruit friends and relatives, saying, Hey, I can do this, you can do this, too. Probably your biggest recruitment is going to be success stories.

On the negative side, if you systematically eliminate and discriminate and harass women out of the trades, then that is what the word is going to be out there on the streets. It's a self-fulfilling situation and it's an insidious thing. It's something you can't prove one way or the other, but a lot of women come in and introduce themselves as a friend of so-and-so that are in the trade or in the apprenticeship program.

Those people know what they're getting into, because they've talked to a real live body who has said, this is what it is, and this is what it's like, and this is what it could be, and this is what you can become, and this is how much money you're going to make, and this is what you're going to learn. Those people are very much eyes wide open, know what they're getting into. When you've got someone who's making thirty, forty, fifty thousand bucks a year living down the street from you, it's a pretty good testimony.

— *Bernadette* —

This is an economic issue as much as it is a gender issue, and if you take that away, you find that people have always protected their territory. I mean, after that first group of women came in, I'm sure the boys went in the back room and said, hey, that's enough of that.

I look back on my decision. My decision was based on, I want to do this for me. I was not thinking about being a pioneer, about being an activist. But I became all of those things because I wanted to do that for me and for my children.

I'm not aware of any good reason why women aren't today at least 15 percent of the workforce in every construction trade. I would certainly take exception to any suggestion that it was the result of a failure of will or ability on the part of tradeswomen.

One surprise to me during interviews with tradeswomen pioneers was the consistency of the response to my question, Did you think this was a good choice for you?

204 — We'll Call You If We Need You

The lack of regret, the affirmation expressed about becoming a tradeswoman—whether people stayed in or left the industry—was a testament to why the fight to open these jobs remains important for reasons beyond economics. Women who had built successful careers were unanimously adamant that lots more women should have that opportunity. I found women consistently generous in trying to explain the likely frame of mind or circumstances of those who placed obstacles in their way. Yet for me, there are stories that, in every stage of work on this manuscript, brought the process to a halt as I became overwhelmed. I may never be able to answer—how could anyone treat Karen that way? or Cynthia? or Paulette? or . . . ? That they are deserving of fair treatment and respect seems unmissable. At the most profound human level, I don't understand parts of this book.

But on a more practical level, several conclusions seem fairly obvious. First, women familiar with the industry and crafts do not find any reason inherent in the work itself that justifies the extraordinarily low percentage of women. Second, because gender is so entrenched as a defining criterion in the construction industry workforce, the elimination of barriers for women requires vigilant external pressure. And, given the high personal cost to tradeswomen involved in the affirmative action efforts originating in the 1970s, relative to the minimal gains accomplished, renewed efforts to allow women their rightful place in the industry should be undertaken only if the intention is a serious one. Meaning: a legitimate career opportunity is being offered, along with access to the training that will lead there, making investment in a three-to-five-year apprenticeship likely to be worthwhile. Meaning: unnecessary obstructions will be cleared from the path by equipped institutions, not lone pioneers.

Meaning, affirmative action works, and more of it works better. Without specific hiring goals for "others," any industry that relies on informal hiring networks will likely keep reproducing itself as—if not "family"—at least, "looks like me/shares my interests and outlook." When there was strong federal monitoring and enforcement, and where federal regulations were reinforced by local ones or by a developer's mandate, affirmative action had positive results for women's employment, and also created a pressure for changes that would make it easier to attract and retain women. Hiring goals should be escalated regularly in order to encourage companies to incorporate women into their core workforce; and to see the full training of female apprentices as a sound business investment. Goals need to be written to ensure employment at journeylevel, so that the revolving door—women working as apprentices but not making a good annual wage at journeylevel—stops spinning women out of the industry.

Meaning, a critical mass of women—that 15 percent—needs to be achieved as quickly as possible so that the "pioneering" phase (which has already lasted twenty years) ends soon. Until a significant number of women are brought in, the message will continue to be: women (with token exceptions) are not capable. Women are entitled to a workplace that is as respectful and comfortable for them and as reflective of their needs and priorities as it is for men. They are entitled to be "settled in."

Which will mean making major adjustments in the organization and culture at all points of the industry. Structural changes, not cosmetic ones. An entryway designed for an 18-year-old living with mom and dad will need to be remodeled for an adult raising children. For example, when contract negotiators lower first-year apprentice pay in exchange for an increase in the journeylevel rate, they work against workforce diversification.

In the twenty years women have been in the industry, tradeswomen individually, in informal small groups, organized within their union local or trade or city, or for brief moments nationally, have offered their ideas and their efforts for increasing the percentage and retention of women. To government monitoring agencies. To union leadership. To contractors. To training programs and vocational schools. To others in the labor movement, in the women's movement. Earnestly. Angrily. Stridently. Diplomatically. To anyone who wanted to listen and to many who did not. It is the perseverance of tradeswomen and their meagerly funded grassroots organizations that has maintained the outpost that has kept the issue of women in the trades alive.

Any renewed intention to make affirmative action successful needs an ongoing policy table that involves working tradeswomen—apprentices committed to making the trades their career and veterans with their hard-won wisdom. And it can't be an all-white seating. We have to get beyond the occasional showcase job that employs women and minorities; regional development plans need to include a commitment to affirmative hiring that is long-range and multi-project. But at only 2 percent of the workforce, tradeswomen lack clout and need allies. Community development groups, women's organizations, other segments of the labor movement, and supporters inside the industry need to pressure for the inclusion of more women, and for the programmatic funding needed not just to recruit them, but to ensure that they can build successful careers.

— Nancy —

A lot of jobs are working four tens, they're working from six o'clock in the morning until five. Or six-thirty until five. The union management needs to address to the contractors accommodations for not only women—there's getting to be more and more men that have custody of their children. I think childcare's a big deal. I think that it is a union problem. It is an industry problem. It is a national problem. What we're basically saying, well, that's the woman's problem—and it's not.

— Randy —

When the Washington State Labor Council sent out all these questionnaires for why apprentices don't make it, the top thing was childcare, from the women. Because we start work at seven o'clock in the morning. If they live in Auburn and they're working in Everett, they're leaving at four-thirty in the morning. A lot of daycare centers aren't open.

— Kathy —

I had a lot of problems dealing with abandoning my children for the sake of my apprenticeship. I think that's a real problem with construction. Obviously, if you go to work at the same place and the same time every day, they can facilitate having a daycare there at that location for toddlers, sick kids, or whatever. Working in construction, you might be on this jobsite this day and another jobsite the next day.

I give the carpenters here locally credit, because they have switched to daytime [apprenticeship] classes—not to accommodate mothers and single parents—but I'm sure it's much easier. Instead of going two or three nights a week for four hours, each quarter they go to class for a week for 40 hours. And I'm sure it's much more effective as far as training and much more easier for any single-parent-type situations. I would recommend that. It's hard enough being gone, working eight hours a day and maybe driving a good three or more hours a day—and then going to school all night long.

— Gay —

In the matter of training, after you got basic training, everything was how skillful you were at getting information on the job. I think that training should be an ongoing process.

There should be time for it, always, when you're on the job, depending on what situation you come from, be it apprentices or journeymen. If there's something going on that a person doesn't know, there should be a time where they can take time to learn that. It shouldn't have to be a thing where you basically have to sneak around and hope to hell someone will give you the information.

— Cheryl —

I think if people were educated more to know that individuals—be they black, white, female, whatever—are capable. . . . You can always see the ones that are basically going to become foremen later on down the line, because they're groomed for those positions. As apprentices, basically, they start them off at the prints all the time, with, These prints are for this and that, and This is how you scale it. Everything is made easy for them, so it's a easy transition to go from worker to foreman. What I think needs to happen, while this grooming process is going on, that it will become more inclusive. Not just picking out one or two people that you necessarily like, or that you see something in. Extend it. Extend it to minorities and to that female and—everyone.

— Diane —

Currently, in our local, in order to be a foreman you need to take a foreman's class, and they go through some training—ten weeks, one night a week, a couple hours a night. They cover scheduling and planning, and there's a section in there on chemical dependency and how to recognize it and how to deal with it. And harassment, racial or sexual harassment, and how to document it. Anybody can take it, so they definitely don't exclude you.

When the women go through this foreman training class and employers do call the hall and say, I need some foremen, the hall does have the opportunity to refer and recommend individuals.

Although many of the tradeswomen I interviewed might reasonably have left the trade union movement in bitterness, I found instead a general willingness to be hopeful that some day.... After idealism, naiveté, and sentimentality have faded, there are still two practical reasons to conclude that tradeswomen's best potential allies within the industry are the construction unions.

First, if women were to become, say, 25 percent of the construction workforce (making it no longer a "nontraditional" field), the victory would be worth much less if—as happens so often when gender shifts in an occupation—the wage package and safety conditions, won and maintained by unions, were simultaneously depressed. The unions also have a self-interest in the diversification of their memberships; to the extent that they are perceived to be exclusive clubs rather than open and inclusive labor organizations, their claim to public work is compromised.

Second, to tackle the gender barriers in the industry involves solving a set of discrete but interconnected problems that can best be addressed when labor is organized. International unions have the budgets and organizational capability and experience to create educational tools, replicate successful programs, conduct long-term studies, lobby for legislation, and negotiate contract language. All apprentices in their final year, for example, could be trained to be foremen—both diversifying the foreman pool and ensuring that all new journeylevel mechanics are trained to interrupt harassment. Union locals could combine resources to establish centralized childcare centers that would accommodate construction hours. Or maybe a local would decide to negotiate a change in work hours to match daycare center schedules. Unions could take on the contractors who refuse to hire or consistently mistreat women.

Construction workers are by necessity resourceful and flexible thinkers. When blueprints are inaccurate, or when the tools and material ordered are not the ones that arrive, the job must still be done. Quite often it is up to the individual mechanic at the site, who can see the conditions and obstacles, to decide the details of how best to accomplish the job for maximum quality and efficiency. This same intelligence and creativity needs to be drawn on to re-think the ground rules and traditions for training, hiring, and layoff, so they cannot serve to mask discrimination. Rather than quibble over what percentage of an apprentice's hazing is due to their gender or race and what to just being an apprentice—end the "tradition" of hazing altogether. Print up and distribute a Handbook of Apprentice Rights to all apprentices that clarifies expectations for skills training, on-the-job behavior, sanitation facilities, and procedures for handling problems.

Unions provide a ready-made structure of stewards, business agents, and training directors. This structure needs to be augmented with communicated procedures to follow when those don't work—just as a transfer switch kicks a hospital's electrical system over to emergency power when the regular system fails. New female apprentices in their probationary period especially need this protection and support. Increased reporting of harassment and discrimination will be a healthy sign, as it has been for women in the military, that women's numbers have increased enough to embolden them out of silence. Leadership needs to open up. These power shifts will not be comfortable.

Unions will play a key role in the inclusion of women in the industry—either expediting or impeding the process.

— Gloria —

They should treat you with respect, even if they don't like the fact that you're there.

I internalized a lot of that stuff, and I didn't say a lot of things to the business agents, and these business agents need to be made aware. There should be maybe a more direct line from the women to the business agents for serious problems. You shouldn't have to go through the foreman, through the contractor, if it's something that you really need to have discussed or taken care of. Because when the guys on the job find out about it, you're ostracized. And you got to work with those guys. So there should be a way of having to take care of those things without getting everybody involved in it, and without everybody knowing about it.

— Cheryl —

I think if there was a organization installed within our union—what we've tried to implement into our union, which is Minority Affairs—it would be a vehicle for women to outreach to. For someone to be a visual image people can see—if you have problems, get in touch with this person or that person and they can help you. If there's someone you can relate to on that level—*Yeah, I know that she's been through this or he's been through this, and I can talk to them and they can give me firsthand information on what's going on*—then they would feel more comfortable. 'Cause I know I would have felt more comfortable if I had've had someone that I could have talked to on different planes. But I felt like I was out there basically doing everything as if it was all brand-new. I was a pioneer with things, and it taught me a lot. But I would like to be able to use the experience that I've gained to help other people that are going through the same thing.

Any problems you have, we'll be like an outreach program, or we can direct them to someone that can help them, if it's the president or the E Board or whatever. 'Cause I know as a first-year, second-year, third-year apprentice, I wasn't aware of the different organizations in our local union and what their functions were. And I think if there was a particular organization, someone that was empathetic towards the minority members or members with special problems or special needs, then they would be more willing to talk to them.

— *Paulette* —

What needs to happen, I think first, is that they need to have comprehensive sensitivity training for men. What I've experienced is that the individual men look at it differently when they're by themselves and it's just you and them. The group dynamics are what are the problems. They're afraid of bucking the group attitude. I think if you had some sort of comprehensive sensitivity training around women being in this space, on this jobsite, in this trade, and belonging there—if they were trained as a group to look at it a certain way, there wouldn't be all these attitudes about it. I mean, you're invading, you're taking their last refuge—you've got all these other things you can do, why are you over here with us?

I don't even know if most of what they feel and do is out of malice. I think they really don't understand. And they really believe you shouldn't be here. The everyday guy out there on all those everyday jobs is the one you've got to target. If the training doesn't get to them, you can forget it, 'cause that's who the women got to deal with. Who can't even tell you why they hate you. They don't even know. I think, yeah, that's the first thing.

— *Karen* —

As foreman, I never got mad at anybody in front of anybody. That's what I would change. Women and blacks—and I shouldn't just say women and blacks, I should say all minorities—usually are treated like dirt in front of the rest of the crew. I watched that just last week. One of the apprentices that is in the program called me, because the business agent that he has cannot relate to him. He considers himself to be a Mexican-American, that's what he wants to be called. And he's getting treated like shit. He called me and just asked me if I would come out to the jobsite and just stand, because he wasn't sure if it was just him imagining these things or if it really was happening.

— Melinda —

I'll tell you another thing which is a major thing to me, is the fact that our card still says *Journeyman*. A journeyman is a man who's willing to journey. I only fit half of that. I'm willing to journey but I'm not a man. And they should acknowledge the fact that there are women in this industry, by either calling you a journeyperson or a journeywoman. It's only a couple of little typesets that they have to change on there to get the cards printed with journeywoman. I still get "Dear Sir" letters and "Dear Brother."

Fourteen years, *Hello,* you know. It's not that they don't know we're there. They don't want to accept it.

———————

When I called up Lorraine Bertosa to arrange the first interview for what became this book, she suggested that she probably wasn't someone I'd want to talk to. She explained that she couldn't be 100 percent positive about her experiences and wasn't even sure she'd encourage other women to come into the industry—so why didn't I call someone else.

Her feelings turned out to be pretty consistent with others I interviewed. As "character-building" as it was to be "pioneers of the industry," the resistance to accepting women as equal partners has taken a physical and emotional toll that is not at all glamorous. I've come to suspect that the suicide rate from the early days (two in my apprenticeship program—one man of color, one white woman) would surprise even those of us who were there.

Tradeswomen with years of experience still struggle for respect. When I asked whether they'd encourage women to enter the trades, the consensus seemed to be a resounding ambivalence. Or a cautionary—only if it's done differently.

But tugging at the other hand are the many positives that make even an adamant "I don't think I could advise a young woman into this industry today" turn into "I guess I would then," when asked, What if there were a lot of women coming in and the industry changed? Because the pay and benefits, the craft and camaraderie— and the chance to add your signature to the skyline—are all still compelling.

As someone who used to advise women going to apprenticeship interviews on the "good answers" to give to improve their chances for acceptance, I hope now for a new surge of interest among young women, who would instead be walking into their interviews with good questions for industry gatekeepers.

— Gay —

You have got to empower women and organize them before they can come to a place where working in the trades will really be attractive. One of the reasons I don't think women come in, and why women don't stay, is because they're so insecure about themselves. And then we put them into a situation that's really hard and it only gets worse. But I think we have to start earlier and make women realize, *There is nothing out there you can't do.*

But we have to be able to organize ourselves as construction women, and get by this fear that, If I do something, there can be retribution. What we have to make them realize is, if there are 300 of us or 200 of us and something happens, then there is no retribution.

I think what we have an obligation to do is to help get rid of some of that fear, to really empower younger women and encourage them to be part of this group, and, when you're going through your training, if something happens, immediately call someone and say, "Such-and-such a thing is going on, how do I deal with it?" But we have to become a voice, a bigger, stronger voice.

— Helen —

It's always easier to have people in front of you to follow and people behind you that you're pulling along. There's a lot of strength in that. You're not carving out a new history for women. You're not necessarily the only one carrying that banner. When you see a sister on the job, that's worth a lot. Or if you see a sister on a jackhammer on the street, that's very strengthening.

— Lorraine —

And just letting young girls build houses. They just have a ball doing it!

Who's Where

Lorraine Bertosa returned to union construction in 1996, mostly for financial reasons. She had taken a six-year break to reach for more opportunities in leadership. For four of those years she was the building super for a church, managing a $100,000 maintenance budget. She also planned and implemented training programs for women interested in the trades and found she enjoyed teaching.

MaryAnn Cloherty left the trades in 1980, and then resumed her apprenticeship in 1987 with a different local, one where she found more support. Active in her union and the larger labor community in Boston, she teaches evenings in a union-sponsored pre-apprenticeship program and works days as a journeywoman carpenter.

Cheryl Lynn Camp began the electrical trade in July '79 and was the first woman to complete IBEW Local 38/Cleveland's apprenticeship program. She has earned her fire alarm, inspector's, and contractor's licenses. She has held the position of foreman and steward.

Sara Driscoll began her electrical apprenticeship in 1978, when she was 30. She worked in the trade for ten years. Since 1989 she has been a massage therapist. As of 1992, she also became a certified practitioner of the Trager® Approach, a form of movement re-education.

Gloria Flowers was raised, one of six children, in a blue-collar family that stressed a work ethic and self-sufficiency. After working seven years with the tools, including her apprenticeship, she moved into her current position as plumbing inspector.

Bernadette Gross has worked for Wider Opportunities for Women (WOW) in Washington, D.C., for four years. She is the Non-Traditional Skills Instructor, teaching safe use of hand and power tools for the local Work Skills Program, a training program for low-income women. She also works with WOW on national programs addressing women's economic issues of childcare, transportation, and economic self-sufficiency.

Barbara Henry transferred her union card from a construction local to a utility local in 1994, following a downturn in construction in the D.C. area. She is currently employed by Potomac Electric Power Co. as an electrical mechanic, working with high voltage (4 kv-138 kv) at the generating end of electrical power. She's gotten her commercial driver's license, is working on her master's, and just bought a house.

Melinda Hernandez will always be a Journeywoman, though due to a lung condition, she can no longer work with her tools. She is employed as an Eligibility Specialist by the Human Resource Administration in New York City, interviewing people who are in need of Medicaid.

Paulette Jourdan, after leaving union construction, has worked as a self-employed plumber since 1988. She remodels and repairs homes and small businesses. As a board member of Women Empowering Women, she was actively involved in EnTrade, a training and support program for women interested in entering the trades.

Donna Levitt began her apprenticeship in 1980, never imagining she would work for the union. Currently she is the senior Business Representative for Carpenters Local 22 in San Francisco, the only woman heading a carpenter construction local in the U.S. She hopes to make a difference.

Cynthia Long has worked with the tools as an electrician for over nineteen years, including two years as electrical superintendent on the World Financial Center in New York City. She has always taken an active role as an advocate for tradeswomen, present and future. She has taught at Cornell's Trade Union Women's Studies Program and at the Northeast Summer School for Trade Union Women.

Randy Loomans worked eleven continuous years as an ironworker. In 1993 she was hired as a full-time construction trades instructor at Renton Technical College for the Apprenticeship and Non-Traditional Employment for Women (ANEW) Program. She is now employed by the Washington State Labor Council as Education Director, representing labor's interests and involvement in apprenticeship, school-to-work, and welfare-to-work.

Nancy Mason was Training Director of Local 46 IBEW in Seattle for seven years. She resigned in 1996 and returned to working with the tools as a maintenance electrician at Sea-Tac International Airport. She is currently employed as Apprenticeship Program Manager for the state of Washington. She serves on the executive board of her local and chairs the Joint Apprenticeship Training Committee.

Diane Maurer completed her apprenticeship in 1983. She worked as a foreman, general foreman, and project manager from 1988 until 1991, and from 1991 until 1995 taught apprentices. She took a hiatus from the tools in 1992, and currently works as an at-home mom with two young children, considering various options for future employment including design, inspection—and returning to electrical construction.

Mary Michels has been in the trades since 1980. Being an ironworker has given her financial freedom and work she enjoys.

Karen Pollak was carpenter superintendent on construction of the I-90 floating bridge, the first time a project in Seattle had two women in foremen positions. For three years she worked for the Port of Seattle as a maintenance carpenter in the marine division. Disabled from trades work by a car accident, she then taught in technical and community college pre-apprenticeship training programs for six years. Currently, she is on pause.

Maura Russell is owner-operator of Pipelines, Inc., a plumbing and heating company in business since 1983 and currently employing five women. As a licensed master plumber, she has had the opportunity to train a number of women over the years, some of whom have gone on to have their own companies.

Irene Soloway has been in the trades since 1979. For the past twelve years she has been a carpenter at Lincoln Hospital in the South Bronx. For ten of those years she's been a union steward, including serving as General Steward for all the carpenters in the Health and Hospitals Corporation. Three years ago she began going to school at night, taking science courses in preparation for a new career in health care.

Diana Suckiel is currently on a leave of absence from plumbing to stay home with her teenage son. She is proud to have gone through her apprenticeship in "the rainbow class": one Native American, one Mexican American, one African American man, three African American women, two Caucasian women, the rest, Caucasian men.

Angela Summer was a journeylevel plumber for ten years. She then worked for three years as a Safety Compliance Officer for the State of Washington Department of Labor and Industries, enforcing the WISHA ACT, a state program governed by OSHA. Currently she's employed as a Safety Consultant for Labor and Industries.

Mercedes Tompkins, after she left the trades, spent nine years as director of Casa Myrna Vasquez, a Boston agency using a comprehensive multi-disciplinary approach to address issues of violence for women, children, and families. She is currently manager of the campaigns unit at OXFAM, a non-profit organization that provides funding for grassroots self-help solutions to poverty and hunger in the global south.

Barbara Trees currently works as an exhibit carpenter at the New York Javits Center. During the past sixteen years she has run for local and district council office, served as a shop steward in her local, organized the District Council Women Carpenters Committee, and founded New York Tradeswomen.

Yvonne Valles found that her ten years in the painting trade opened doors and improved her lifestyle financially. She has been employed for the past two years as a correctional peace officer at California Institution for Men at Chino.

Helen Vozenilek served her apprenticeship in Albuquerque, New Mexico, and later worked for the Department of Parks and Recreation in San Francisco. In 1997, after working as an electrician for sixteen years, she entered the Graduate School of Journalism at University of California/Berkeley, a two-year full-time program.

Kathy Walsh after several years as a journeylevel carpenter, became a recruitment officer for the Hometown Plan in Kansas City. She then joined with another woman carpenter who had started her own construction company. They received a grant for pre-apprenticeship training, and helped about thirty women and minority men get started in apprenticeships. After the company closed for financial reasons, Kathy found an interior finish contractor who is training her as an estimator.

Gayann Wilkinson loved working as an ironworker, though it took its mental and physical toll. She became an inspector and then a bridge engineer, overseeing million dollar projects. A founding co-chair of the Boston Tradeswomen's Network, her activism led to an appointment as Director of Apprentice Training for the Commonwealth of Massachusetts.

Deb Williams, a tradeswoman for twenty years, still remembers painting the Mystic River Bridge when she was 18 years old. She is currently a general foreman for the Massachusetts Bay Transit Authority. She is Recording Secretary of Local 1138 and on The Women's Committee of her Painters District Council.

Marge Wood coordinates apprentice related instruction for the Wisconsin Technical College System. Her credibility as the Apprenticeship Consultant is built on twelve years of experience as the first woman plumber in Madison, Wisconsin, and an enduring belief in the value of apprenticeship training.

Record of Interviews

Lorraine Bertosa. November 30, 1991.

Cheryl Camp. November 16, 1992.

MaryAnn Cloherty. April 13, 1992.

Sara Driscoll. March 23, 1992.

Gloria Flowers. November 29, 1992.

Bernadette Gross. June 27, 1994.

Barbara Henry. June 26, 1994.

Melinda Hernandez. February 27, 1993.

Paulette Jourdan. February 23, 1994.

Donna Levitt. February 25, 1994.

Cynthia Long, February 18, 1992.

Randy Loomans. May 16, 1992.

Nancy Mason. July 13, 1992.

Diane Maurer. May 13, 1992.

Mary Michels. February 20, 1994.

Karen Pollak. July 15, 1992.

Maura Russell. December 23, 1991.

Irene Soloway. February 19, 1992.

Diana Suckiel. April 17, 1994.

Angela Summer. May 13, 1992.

Mercedes Tompkins. March 24, 1992.

Barbara Trees. February 17, 1992.

Yvonne Valles. February 21, 1994.

Helen Vozenilek. February 26, 1994.

Katherine Walsh. April 18, 1994.

Gayann Wilkinson. April 6, 1992.

Debra C. Williams. March 25, 1992.

Marge Wood. April 16, 1994.